Fundamentals of
Elementary Mathematics:
Geometry

Fundamentals of
Elementary Mathematics:
Geometry

Merlyn J. Behr and Dale G. Jungst
Northern Illinois University

DISCARDED

Academic Press

New York and London

ACADEMIC PRESS, INC.
111 Fifth Avenue, New York, New York 10003

United Kingdom Edition published by
ACADEMIC PRESS, INC. (LONDON) LTD.
24/28 Oval Road, London NW1 7DD

LIBRARY OF CONGRESS CATALOG CARD NUMBER: 77-182621

PRINTED IN THE UNITED STATES OF AMERICA

To our respective families

Joan	*Reva*
Cynthia	*Steve*
Craig	*Carmen*
	Roger
	Joanne

contents

3

subsets of lines

4

distance, measure, congruence of line segments and angles

5

curves

6

congruence and similarity

preface

We have tried to meet the need for a geometry book which combines an intuitive introduction to the concepts with a careful and precise statement of them. The topics are developed in such a way that intuitive motivation leads to an in-depth treatment of appropriate topics; in this sense it is consistent with its predecessor, "Fundamentals of Elementary Mathematics: Number Systems and Algebra."

The intended audience of this book is prospective and inservice elementary and junior high school teachers. Therefore, in the selection of topics we were guided by the CUPM I (1961) recommendations, the content of the present elementary and junior high school curricula, and some contemporary mathematics curriculum projects. The level of treatment exemplifies our personal philosophy that elementary school teachers must know the subject of geometry well beyond the level of sophistication at which they teach. In this we concur with the following suggestion given in the CUPM I (1961) recommendations:

The material in this course* might in a sense, duplicate material studied in high school by the prospective teacher, but we urge that

* Changed to singular by the author.

*this material be studied again, this time from a more sophisticated
college-level point of view.**

Since the language, symbolism, and geometric models used in
this book are compatible with present elementary and junior high
school textbooks, the future teacher can easily make the transition
from the study of geometry using this book to his teaching of
geometry.

A prepublication draft of the manuscript for this book was used
successfully with a group of inservice elementary and junior high
school teachers whose mathematical preparation at the beginning
of the course varied from not more than one year of high school
algebra to a more adequate background. Although we recommend
at least one semester of college mathematics dealing with sets,
logic, and number systems from a book such as "Fundamentals
of Elementary Mathematics: Number Systems and Algebra" as
prerequisite for this book, our experience with the prepublication
draft suggests that it can be used successfully with students having
not more than one year of high school algebra. Indeed, the book
was written to be self-contained in the sense that the topics associated with sets, relations, functions, logic, and the real number
system, which are necessary for the study of the geometric development, are treated within the book. If the student has had the
suggested prerequisite work, these sections can serve as review.

This book was written to provide adequate material for a one-
semester course. For students who have not had prior study of
sets, logic, and number systems, Chapters 1–8 will constitute sufficient material for a one-semester course; students who have the
suggested prerequisite can probably complete Chapters 2–10 in
a semester course.

The several chapters in this book were included to fulfill general
objectives as follows: Chapter 1, to give background material for
the study of geometry; Chapter 2, to emphasize the structure of
mathematics through a finite geometry which also demonstrates
how different assumptions about the undefined concept of point

---

* Mathematical Association of America, "Recommendations for the Training of
Teachers of Mathematics: A Summary," 1961.

can lead to a very different notion of line and other geometric objects than that to which students are accustomed, and to take students through the beginning of a formal development of Euclidean geometry to demonstrate how it can be developed from a small list of assumptions; Chapters 3–8, to present the many geometric concepts which we consider important for the intended audience (although these chapters are logically organized, they are less formal than Chapter 2); Chapter 9, to provide the necessary prerequisite material for Chapter 10; Chapter 10, to present an introduction to Euclidean geometry via transformations.

acknowledgments

To Joan Behr we offer many thanks for her competent work as the manuscript typist. The Mathematics Department at Northern Illinois University provided services for duplication of the pre-publication draft used in classroom testing; for this we are thankful. The patience and cooperation of the students who studied from this duplicated work is acknowledged and appreciated.

We are especially grateful to the editors and staff of Academic Press with whom we continue to find working both a privilege and a pleasure. We acknowledge the helpful criticisms and suggestions given by Professor Jo Phillips, University of Cincinnati, and Professor Carroll O. Wilde, U.S. Naval Postgraduate School, who again served as very competent and conscientious reviewers of the manuscript.

Our warm appreciation goes to our families for their patience and understanding during the development of this work.

We accept full responsibility for the contents of this work; for whatever in it is good we jointly accept credit and for whatever is bad we each cheerfully blame the other. We invite all readers of this book to send us their criticisms, and we will appreciate having any errors called to our attention.

glossary of symbols and notation

{ }	braces (commonly used to denote sets)
$\in$	is an element of
$\subseteq$	is a subset of
$\subset$	is a proper subset of
$\approx$	is equivalent to
$\cup$	union
$\cap$	intersection
$\times$	cartesian or cross product
$\sim$	complement
$A \setminus B$	relative complement of B with respect to A

RELATIONS AND FUNCTIONS

$x R y$	x is related to y by R
$F(x)$	the F-image of x; the value of F at x, F of x
$\left. \begin{array}{l} F(x) = y \\ F\!:\!x \rightarrow y \\ x \xrightarrow{F} y \end{array} \right\}$	F of x equals y; x gets mapped to y by F; the F-image of x is y
$\circ$	composition of transformations

GLOSSARY OF SYMBOLS AND NOTATION

REAL NUMBERS

$\mathbb{R}$	the set of real numbers
$a, b, c, \ldots$	elements in $\mathbb{R}$
$<$	is less than
$>$	is greater than
$\leqslant$	is less than or equal to
$\geqslant$	is greater than or equal to
$\doteq$	is approximately equal to

GEOMETRIC FIGURES

S	space in Euclidean geometry
$A, B, C, \ldots$	points in S
$\mathscr{L}, \mathscr{L}_1, \mathscr{L}_2, \ldots$ $\mathscr{M}, \mathscr{M}_1, \mathscr{M}_2, \ldots$ $\mathscr{N}, \mathscr{N}_1, \mathscr{N}_2, \ldots$	lines in S
$\alpha, \beta, \gamma, \ldots$	planes in S
$\mathscr{C}, \mathscr{C}_1, \mathscr{C}_2, \ldots$	curves in S
$\mathscr{P}, \mathscr{P}_1, \mathscr{P}_2, \ldots$	polygons in S
$\mathscr{F}, \mathscr{F}_1, \mathscr{F}_2, \ldots$	surfaces in S
$\mathscr{S}, \mathscr{S}_1, \mathscr{S}_2, \ldots$	solids in S
$\mathscr{A}, \mathscr{B}, \mathscr{D}, \ldots$	arbitrary subsets of S
$\overleftrightarrow{AB}$	line AB; line which contains points A and B; line determined by points A and B
$\overline{AB}$	line segment AB; line segment determined by points A and B
$\overset{\circ}{\underset{}{\rightarrow}}{AB}$	half-line AB; half-line determined by points A and B
$\overrightarrow{AB}$	ray AB; ray determined by points A and B; ray with endpoint A and containing point B
$\overleftrightarrow{AB}{:}P$ ($\mathscr{L}{:}P$)	the P-side of line AB (line $\mathscr{L}$); the half-plane which contains P
$\angle ABC$	angle ABC; the angle determined by points A, B, and C with vertex B
$\triangle ABC$	triangle ABC; the triangle determined by points A, B, and C
$\square ABCD, \square ABCD,$ $\square ABCD, \square ABCD$	square, rectangle, trapezoid, parallelogram $ABCD$; the square, rectangle, trapezoid, parallelogram determined by points A, B, C, and D
Int( )	interior of
Ext( )	exterior of

Reg($\mathscr{C}$)	region determined by simple closed curve $\mathscr{C}$
▲ABC	triangular region ABC; triangular region determined by $\triangle ABC$
■$ABCD$, ◪$ABCD$, ◣$ABCD$, ▬$ABCD$	the region determined by square, rectangle, trapezoid, parallelogram $ABCD$

RELATIONS AND FUNCTIONS ON GEOMETRIC FIGURES

A–B–C	point B is between points A and C
$m_U(A,B)$	the measure of the distance from A to B with respect to unit U
$\cong$	is congruent to
$\perp$	is perpendicular to
$\parallel$	is parallel to
$\leftrightarrow$	corresponds to
$\sim$	is similar to

1

sets, relations, functions, and logic

Our approach to the study of geometry makes it necessary for the reader to have well established in his mind basic concepts about the nature of mathematics and certain concepts related to the topics of sets, relations, and functions, and also some basic ideas from logic. It is expected that many of these concepts, which we present in this chapter, will be a review for the reader.

1.1 THE DEVELOPMENT OF MATHEMATICS

The modern approach to the study of mathematics is to consider it as a system or structure, or as a collection of substructures. Four essential components are identified in a mathematical or logical system, or structure; they are:

1. undefined terms,
2. definitions (or defined terms),
3. postulates, and
4. theorems.

This list is often said to define "the structure of mathematics."

One may well ask whether undefined terms are necessary. The affirmative answer results from the fact that it is impossible to define every concept in terms of more elementary concepts without arriving at circular definitions. This can easily be demonstrated by referring to a dictionary. If the meaning of word A is sought, word B might be given as a synonym. If the definition of B is found, word C might appear as a synonym. If we keep up this process of finding synonyms, we will eventually run out of different words as synonyms and finally arrive back at word A, and therefore, have word A defined in terms of word A.

For our study, one of the undefined concepts is that of *set*. "Set" is a mathematical term which is a synonym for "collection" or "aggregate." Undefined terms serve the useful purpose of giving a beginning point from which to begin defining other concepts.

We believe the notion of "defined terms" or "definitions" is familiar and needs no further comment.

The role of postulates is fundamental in a mathematical system. *Postulates* are assumptions which provide the foundation for the structure. The role and the use of postulates are not unique to mathematics. Indeed, our system of government is based on some assumptions. Our forefathers based the United States Constitution on a premise, among others, that all men are created equal. We make note of two things about this premise:

1. It is indeed an assumption.
2. No attempt is made to establish the truth of (prove) this statement.

From the postulates other facts, called *theorems*, are deduced. In our system of government we have a collection of theorems; we call them "laws" or "statutes." For example, many Federal statutes relative to civil rights establish truths which are deducible from our assumption that all men are created equal.

The question of why we do not prove everything in mathematics (that is, why make assumptions?) may have occurred to the reader. The answer to this is similar to the explanation for the need for undefined terms. If we start without assumptions, then there is nothing in the system from which to establish theorems.

As a fundamental concept of our entire study, we accept the notion of equality to be as follows: to write $x = y$ means that x and y are names for the same object. This makes the notion of equality very general. For example, if A and B name sets, to write

$A = B$

means that A and B are names for the same set.   If x and y name numbers, to write

$x = y$

means that x and y are names for the same number.   Similarly, when we write

$5 + 9 = 14$

we mean "$5 + 9$" and "14" are names for the same number.
   We make the following assumptions about equality:

1.   *The Reflexive Property.*   For every x, $x = x$.
2.   *The Symmetric Property.*   For all x and y, if $x = y$, then $y = x$.
3.   *The Transitive Property.*   For all x, y, and z, if $x = y$ and $y = z$, then $x = z$.

We also assume:

4.   *The Substitution Principle.*   For every x and y, if $x = y$, then "x" and "y" may each be substituted for the other in a statement.

1.2 THE SET CONCEPT AND SET NOTATION

We have already indicated that "set" is an undefined word which is a synonym for "collection," "aggregate," and some other words. Those objects which make up the collection which we wish to consider a set are called *members* or *elements* of the set.   If "A" is the name for a set and "x" is the name for an element of A, we indicate this by writing

$x \in A$.

The statement "$x \in A$" is read "x is an element of A," "x is a member of A," "x belongs to A," or "A contains x as a member."
   When we wish to communicate with others about a particular set, it is necessary to describe the set so that others will know precisely what objects we have in mind.   To do this it is necessary and sufficient to give a criterion for determining whether any given object belongs to the set.   When such a criterion is given, the set is said to be *well defined.*   A set can be well defined in at least

three ways:

1. giving a verbal description of the criterion for membership,
2. listing the elements and enclosing the list in set brackets, or
3. using set builder notation.

The three methods of describing or defining a set are illustrated in Example 1.1.

EXAMPLE 1.1   Other descriptions of

$\{x|x$ is a whole number and $5 < x < 45\}$

are:

a. the set of all whole numbers between 5 and 45, and

b. $\{6,7,8, \ . \ . \ . \ ,44\}$.

It is essential that the reader clarify in his mind the distinction between an object x and the set whose only element is that object $\{x\}$.   This distinction is analogous to that between an apple and an apple in a basket.   The symbol "x" names an object, whereas, "$\{x\}$" names the set whose only element is named "x."

Since we have made no restrictions on what objects can be members of sets, we have allowed the possibility that sets can be elements of sets.   The ideas are illustrated in Example 1.2.

EXAMPLE 1.2   If $A = \{a,\{a\},\{b\}\}$, the following are true statements:

a. $a \in A$.
b. $\{a\} \in A$.
c. $b \notin A$ ($\notin$ means "is not an element of").
d. $\{b\} \in A$.

No restriction is made concerning the number of elements which must be contained in a set.   Indeed, we allow a collection of no objects to be called a set.   It is called the *empty set* and is denoted by $\varnothing$.   Other words sometimes used for "empty set" are, "null set" or "void set."   A set which contains at least all of the elements which belong to every set in a given discussion is called a *universal set;* we will denote a universal set by U.

Exercise set 1.1

1. Give verbal descriptions of the following sets:

    a. {1,2,3,4,5}
    b. {x|x is a counting number}
    c. {1,2,3, . . . ,105}
    d. {x|x is a whole number and x > 3}
    e. {x|x is a whole number and 3 < x < 10}

2. Give another description of the following sets:

    a. The set of five-eared bullfrogs residing in your house
    b. The set of letters in the English alphabet between a and g
    c. {x|x is a whole number and 5 < x < 8}

3. Use set builder notation to describe the following sets:

    a. {10,20,30, . . .}
    b. The set of whole numbers greater than 5 and less than 10
    c. {5,6,7,8,9,10}

1.3 RELATIONS ON SETS

Throughout this chapter we will use upper case letters $A, B, C, \ldots$ as names for sets. Consistent with our fundamental concept of equality we have:

For all sets A and B, $A = B$ if and only if A and B have exactly the same elements.

Equality is a way to compare sets; we call it a *relation* on sets. Another very important way to compare sets involves the concept of one-to-one correspondence which is given in Definition 1.1.

Definition 1.1 A *one-to-one correspondence* between two sets A and B is a pairing of elements of A and B such that with every element of A exactly one element of B is paired, and with every element of B exactly one element of A is paired.

If there exists a one-to-one correspondence between two sets A and B, we call them *equivalent* sets and write $A \approx B$. The concept of one-to-one correspondence is illustrated in Example 1.3.

EXAMPLE 1.3 If $P = \{a,b,c,d\}$ and $Q = \{1,2,3,4\}$, then the pairing

$$a \longleftrightarrow 1, \qquad b \longleftrightarrow 2, \qquad c \longleftrightarrow 3, \qquad d \longleftrightarrow 4$$

is a one-to-one correspondence between P and Q. The pairing

$$a \longleftrightarrow 2, \qquad b \longleftrightarrow 1, \qquad c \longleftrightarrow 4, \qquad d \longleftrightarrow 3$$

is another one-to-one correspondence between P and Q and can be diagrammed as shown in Fig. 1.1.

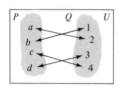

FIG. 1.1
A one-to-one correspondence.

The following properties of the relation $\approx$ on sets can be proved but we omit the proof here and simply state the results.

1. *The Reflexive Property.* For every set A, $A \approx A$.
2. *The Symmetric Property.* For all sets A and B, if $A \approx B$, then $B \approx A$.
3. *The Transitive Property.* For all sets A, B, and C, if $A \approx B$ and $B \approx C$, then $A \approx C$.

Another relation on sets is the subset relation which is given in Definition 1.2.

Definition 1.2 For all sets A and B, A is a *subset* of B (we write $A \subseteq B$) if and only if every element of A is also an element of B. That is, $A \subseteq B$ if and only if for every $x \in A$, $x \in B$ (or for every x, if $x \in A$, then $x \in B$).

The statement

For every x, if $x \in A$, then $x \in A$

is true; therefore, every set is a subset of itself. Thus when we write $A \subseteq B$, we allow the possibility that $A = B$. Sometimes we wish to indicate that A is a subset of B and exclude the possibility that $A = B$. In this case we call A a *proper subset* of B and write $A \subset B$.

Definition 1.3 For all sets A and B, A is a *proper subset* of B (we write $A \subset B$) if and only if $A \subseteq B$ and $A \neq B$.

We illustrate the relations of equality, subset, and proper subset in Example 1.4.

EXAMPLE 1.4 Let $A = \{a,b,c,d\}$, $B = \{b,a,c,d\}$, $C = \{a,b,c\}$, and $D = \varnothing$. Then the following are true statements:

a. $A = B$.   b. $A \subseteq B$.   c. $D \subseteq B$.
d. $C \subseteq A$.   e. $C \subset A$.

In order to show that for sets A and B, $A \nsubseteq B$, it is necessary to exhibit an element of A which is not an element of B. The impossibility of exhibiting an element x such that $x \in \varnothing$ and $x \notin A$ leads to the conclusion that

For every set A, $\varnothing \subseteq A$.

A connection between the equality and subset relations for sets is given in Postulate 1.1.

Postulate 1.1 (*The axiom of extentionality*) For all sets A and B, $A = B$ if and only if $A \subseteq B$ and $B \subseteq A$.

Postulate 1.1 gives a way to prove that two sets are equal; that is, by proving that each is a subset of the other.

Exercise set 1.2

1. Exhibit a one-to-one correspondence between the sets $A = \{\text{Jack,Mary,Sue}\}$ and $B = \{\text{Bob, June, Jack}\}$.

2. Exhibit a pattern for making a pairing which will suggest that

   $A = \{100,101,102, \ldots ,200\}$

   and

   $B = \{411,412,413, \ldots ,511\}$

   are equivalent.

3. Indicate whether the following statements are true or false:

   a. For all sets A and B, if $A = B$, then $A \approx B$.
   b. For all sets A and B, if $A \approx B$, then $A = B$.

4. If $A = \{1,2,3,4\}$ and $B = \{1,2,3,4,5\}$, indicate whether the following are true or false:

a.	$A \subseteq B$.	**b.**	$A \subset B$.	**c.**	$B \subseteq A$.
d.	$1 \in A$.	**e.**	$3 \in B$.	**f.**	$\{1,2\} \subset B$.
g.	$\{2\} \in B$.	**h.**	$\{1,5\} \subseteq A$.	**i.**	$\{1,5\} \subset A$.
j.	$\{1,5\} \subseteq B$.	**k.**	$\{1,5\} \subset B$.	**l.**	$\varnothing \subseteq A$.
m.	$\{\varnothing\} \subseteq A$.	**n.**	$\varnothing \in A$.	**o.**	$\varnothing \notin B$.
p.	$\varnothing \subseteq \varnothing$.	**q.**	$\varnothing \in \varnothing$.		

5. If $A = \{a,\{a,b,c\},\{a\},b\}$, indicate whether the following are true or false:

a.	$a \in A$.	**b.**	$a \subseteq A$.	**c.**	$\{a\} \subseteq A$.
d.	$\{a\} \in A$.	**e.**	$\{\{a\}\} \subset A$.	**f.**	$\{a,b\} \subseteq A$.
g.	$\{\{a,b,c\}\} \subseteq A$.	**h.**	$c \in A$.	**i.**	$\{\{c\}\} \subseteq A$.
j.	$\{\varnothing\} \subset A$.	**k.**	$\{c\} \in A$.	**l.**	$\varnothing \subseteq A$.
m.	$\varnothing \in A$.				

6. If $A = \{a,\{a\},\varnothing\}$, indicate whether the following are true or false:

a.	$\varnothing \in A$.	**b.**	$\varnothing \subseteq A$.	**c.**	$\varnothing \subset A$.
d.	$\{\varnothing\} \subseteq A$.	**e.**	$\{\varnothing\} \subset A$.	**f.**	$\{\varnothing\} \in A$.

7. List all the subsets of $\varnothing$; of $\{a\}$; of $\{a,b\}$; of $\{a,b,c\}$; of $\{a,b,c,d\}$; and of $\{a,b,c,d,e\}$.

8. If A is a set with eight elements, how many subsets does A have?

9. If A is a set with n elements, how many subsets does A have?

10. Indicate whether the subset relation ($\subseteq$) on sets is:

    a. reflexive   **b.** symmetric   **c.** transitive

11. Indicate whether the proper subset relation ($\subset$) on sets is:

    a. reflexive   **b.** symmetric   **c.** transitive

12. Let $A = \{a,b,c,d,e\}$, $B = \{1,2,3,4,5\}$, and $C = \{v,w,x,y,z\}$.

    a. Diagram (as in Fig. 1.1) a one-to-one correspondence between A and B.

    b. Similarly, diagram a one-to-one correspondence between B and C.

  c. Use the one-to-one correspondences in parts a and b to de-
      termine a one-to-one correspondence between A and C.

      (*Note:* This suggests the procedure for proving that the
      relation $\approx$ has the transitive property.)

1.4 FINITE AND INFINITE SETS

The reader is familiar with the set of numbers $C = \{1,2,3,4, \ldots\}$,
which is often called the *set of counting numbers*, and also the *set
of even counting numbers*, $C_e = \{2,4,6,8, \ldots\}$. We will demonstrate that there is a one-to-one correspondence between C and C_e;
since we cannot completely list the pairing of elements, we will
give a scheme or pattern, and a formula which will indicate the
precise pairing which is to be made.

$$
\begin{array}{ll}
C & C_e \\
1 & \longleftrightarrow 2 \\
2 & \longleftrightarrow 4 \\
3 & \longleftrightarrow 6 \\
4 & \longleftrightarrow ? \\
5 & \longleftrightarrow ? \\
\vdots & \\
21 & \longleftrightarrow ? \\
\vdots & \\
x & \longleftrightarrow ? \\
\vdots & \\
? & \longleftrightarrow k
\end{array}
$$

If the reader completes the pattern, he will demonstrate a one-to-
one correspondence. It seems surprising that a one-to-one correspondence exists between C and C_e in view of the fact that C_e is
a proper subset of C; however, this is true of every infinite set.

Definition 1.4 A set A is called *infinite* if and only if there is a
proper subset B of A such that $B \approx A$.

Definition 1.5 A set B is called *finite* if and only if it is not infinite.

Definitions 1.4 and 1.5 together with Example 1.5 make it clear that a set is finite provided it contains no proper subset to which it is equivalent.

EXAMPLE 1.5   Consider sets $P = \{1,2,3,4,5\}$ and $Q = \{1,2,3,4\}$, where Q is a proper subset of P.   The following pairing suggests that no one-to-one correspondence exists between P and Q.

$$
\begin{array}{ll}
P & Q \\
1 \longleftrightarrow 1 \\
2 \longleftrightarrow 2 \\
3 \longleftrightarrow 3 \\
4 \longleftrightarrow 4 \\
5 & \leftarrow \text{The elements of } Q \text{ have been exhausted before a mate for} \\
& \quad 5 \text{ was obtained.}
\end{array}
$$

Exercise set 1.3

1.   Let $C = \{1,2,3,4, \ . \ . \ .\}$ and $C_0 = \{1,3,5,7, \ . \ . \ .\}$.   Exhibit a scheme and formula which defines a one-to-one correspondence between C and C_0.

2.   Let $C = \{1,2,3,4, \ . \ . \ .\}$ and $M = \{9,10,11,12, \ . \ . \ .\}$.   Exhibit a scheme and a formula which will prove that C and M are equivalent sets.

1.5   A BRIEF INTRODUCTION TO LOGIC

The subject of logic is very extensive; it is therefore not our intention to present an exhaustive study of the subject.   Indeed, we will present only the concepts of logic which are necessary for the purpose of this text.

1.5.1   Sentences and statements

We assume that the reader is familiar with the concept of a sentence.   Our concern in this section will be with special kinds of sentences called *statements*.

Definition 1.6   A *statement* is any sentence which is either true or false, but not both.

We observe that the definition of a statement requires that a

sentence be either true or false in order to be a statement; however, it does not require that there be agreement on whether it is
true or false. Examples of statements are:

1.6 President Lincoln was born in Kentucky.

1.7 The sum of 26 and 4 is 30.

1.8 $26 + 24 = 71$.

Examples of sentences which are not statements are:

1.9 How old are you?

1.10 $x + 4 = 9$.

1.11 He is a scoundrel.

Examples 1.10 and 1.11 are not statements because it is not possible to declare them either true or false until replacements are
made for "x" and "he"; these sentences are examples of open sentences. Whenever we declare that a statement is true, we will say
it has *truth value T*; and whenever we declare a given statement
false, we will say it has *truth value F*. We refer to *T* and *F* as
opposite truth values.

1.5.2 The connectives *and, or, not*

We next discuss the meaning of the logical connectives *and, or,*
and *not*. We will discuss these connectives by indicating the
truth value of statements which are formed from a given statement
(or statements) and one or more of the connectives.
    The two statements

Chicago is in Illinois

and

Washington is the capital of the United States

can be connected by *and* to form the statement

Chicago is in Illinois and Washington is the capital of the United
States.

The formal definition of the connective *and* is given in Definition 1.7.

Definition 1.7   For all statements p and q, the *conjunction* of p and q, denoted by p *and* q, is the statement which is true only when p and q are both true.

A truth table is a convenient method to summarize the definitions of the logical connectives; the truth table for *and* is given in Fig. 1.2.

p	q	p *and* q
T	T	T
T	F	F
F	T	F
F	F	F

FIG. 1.2
Truth table for and.

The connective *or* is frequently used in mathematical and ordinary discourse.   The two statements

$3 + 4 = 7$

and

$7 + 1 = 9$

can be connected by *or* to form the statement

$3 + 4 = 7$ or $7 + 1 = 9$.

The definition of the connective *or* follows and the truth table is given in Fig. 1.3.

Definition 1.8   For all statements p and q, the *disjunction* of p and q, denoted by p *or* q, is the statement which is false only when both p is false and q is false.

p	q	p *or* q
T	T	T
T	F	T
F	T	T
F	F	F

FIG. 1.3
Truth table for or.

In everyday discourse the word *or* has two meanings; the *inclusive or*, as defined in Definition 1.8 permits both statements connected by *or* to be true; however, the *exclusive or* excludes the possibility that both are true.   The statement

John is dead *or* John is alive

illustrates the *exclusive or;* the truth of either of the statements

John is dead

or

John is alive

excludes the possibility of the other being true.   The meaning given to *or* in mathematics is always the *inclusive or*.

An illustration of the four possible truth-value combinations for the connectives *and* and *or* are given in Example 1.12.

EXAMPLE 1.12   Let the following statements have truth values as indicated:

$4 + 5 = 9$.   *(T)* $4 < 7$.   *(T)*
$9 < 3$.   *(F)* $3 = 5$.   *(F)*

a.   The following conjunctions illustrate the four truth-value combinations in the truth table for *and* (Fig. 1.2).

   $4 + 5 = 9$ and $4 < 7$.   *(T)*
   $4 + 5 = 9$ and $9 < 3$.   *(F)*
      $9 < 3$ and $4 < 7$.   *(F)*
      $9 < 3$ and $3 = 5$.   *(F)*

b.   The following disjunctions illustrate the four truth-value combinations in the truth table for *or* (Fig. 1.3).

   $4 + 5 = 9$ or $4 < 7$.   *(T)*
   $4 + 5 = 9$ or $9 < 3$.   *(T)*
      $9 < 3$ or $4 < 7$.   *(T)*
      $9 < 3$ or $3 = 5$.   *(F)*

The *negation* of simple statements such as

President Lincoln was born in Kentucky

and

$3 + 4 = 5$

is easily formed by the grammatic insertion of the word *not* as follows.

President Lincoln was *not* born in Kentucky

and

$3 + 4 \neq 5$.

The definition of the connective *not* follows and the truth table is given in Fig. 1.4.

Definition 1.9 For every statement p, the *negation* of p, denoted by *not* p, is the statement which is true whenever p is false, and is false whenever p is true.

p	*not* p
T	F
F	T

FIG. 1.4
Truth table for **not.**

For statements which involve one or more logical connectives (compound statements), forming the negation is more difficult; we illustrate the way in which the negations of conjunctions and disjunctions are formed: The negation of the *conjunction*

Corn is yellow *and* potatoes are grown in Idaho

is the *disjunction*

Corn is not yellow *or* potatoes are not grown in Idaho.

The negation of the *disjunction*

Bob is tall *or* Betty is not old

is the *conjunction*

Bob is not tall *and* Betty is old.

We show by the truth table in Fig. 1.5 that the negation of "p *and* q" is "(*not* p) *or* (*not* q)."

Since the truth values in the two columns connected by the

	p	q	not p	not q	p and q	(not p) or (not q)
FIG. 1.5	T	T	F	F	T	F
Truth table to	T	F	F	T	F	T
show that the ne-	F	T	T	F	F	T
gation of **p** *and* **q**	F	F	T	T	F	T
is (not p) or (not q).					↑	↑

arrows (Fig. 1.5) are in each row opposite truth values, it follows that the negation of "*p and q*" is "*(not p) or (not q)*." We leave to the reader the problem of constructing a similar truth table to show that the negation of "*p or q*" is "*(not p) and (not q)*.

Exercise set 1.4

1. Indicate whether each of the following is a statement:

   a. Drive safely!
   b. The nucleus of an atom is positively charged.
   c. From here to eternity.
   d. The United States' national debt is $46 billion.
   e. $x + 5 = 9$.
   f. _____ is young.
   g. Her mother is old.

2. Form the conjunction and disjunction of each of the following pairs of statements, and using the assigned truth values indicate the truth value of each conjunction and disjunction:

   a. Bill is old. (*T*)
      People are funny. (*F*)
   b. Roses are red. (*T*)
      Violets are blue. (*T*)
   c. $4 + 3 = 9$. (*F*)
      $7 + 3 = 12$. (*F*)
   d. $13 + 7 = 25$. (*F*)
      $13 + 6 = 19$. (*T*)

3. Write the negation of each conjunction and disjunction which you wrote in Exercise 2 and indicate its truth value.

4. Use a truth table to show that the negation of "*p or q*" is "*(not p) and (not q)*."

1.5.3   The logical connectives *if · · · then, if and only if*

The two statements

Geometry is interesting

and

Constructing a proof is a challenge

can be connected by *"if · · · then"* to form the *conditional statement*

If geometry is interesting, then constructing a proof is a challenge.

In the conditional statement above, the statement

Geometry is interesting

is called the *hypothesis,* and the statement

Constructing a proof is a challenge

is called the *conclusion.*   The definition of the connective *if · · · then* follows and the truth table is given in Fig. 1.6.

Definition 1.10   For all statements p and q, the *conditional statement* (or *conditional*) with p as *hypothesis* and q as *conclusion,* denoted by *if p, then q,* is the statement which is false only when p is true and q is false.

p	q	*if p, then q*
T	T	T
T	F	F
F	T	T
F	F	T

FIG. 1.6
Truth table for
if · · · then.

We call the reader's attention to the last two rows of the truth table for *if · · · then;* these indicate that a conditional is true whenever its hypothesis is false.   The four truth-value combinations for a conditional are illustrated in Example 1.13.

EXAMPLE 1.13   Let the following statements have truth values as indicated.

$4 + 5 = 9.$ (T) $4 < 7.$ (T)
    $9 < 3.$ (F) $3 = 5.$ (F)

Then, the following conditions illustrate the four truth-value combinations in the truth table for *if* · · · *then* (Fig. 1.6).

If $4 + 5 = 9$, then $4 < 7$. (T)
If $4 + 5 = 9$, then $9 < 3$. (F)
If $9 < 3$, then $4 < 7$. (T)
If $9 < 3$, then $3 = 5$. (T)

A conditional with hypothesis p and conclusion q does not always appear in exactly the "*if p, then q*" form. Other forms for a conditional are illustrated in Example 1.14.

EXAMPLE 1.14 The conditional

If lines are straight, *then* planes are flat

can be written as

Planes are flat *if* lines are straight

or as

Lines are straight *only if* planes are flat

or as

Planes are flat *whenever* lines are straight.

For the conditional

If $4 + 5 = 9$, then $9 < 3$

the following conditional

If $9 < 3$, then $4 + 5 = 9$

is called its *converse*.

Definition 1.11 For all statements p and q, the *converse* of the conditional *if p, then q* is the conditional *if q, then p*.

We note, using the truth values as assigned in Example 1.8, that the conditional

If $4 + 5 = 9$, then $9 < 3$

has truth value F; whereas, its converse

If $9 < 3$, then $4 + 5 = 9$

has truth value T; that is, the truth value of the converse of a given conditional cannot be determined from the truth value of the given conditional.   In particular, the fact that a given conditional is true does not guarantee that its converse is true.   This can be seen from the truth table in Fig. 1.7.   In the two rows shaded the conditional *if p, then q* and its converse *if q, then p* have opposite truth values.

p	q	*if p, then q*	*if q, then p*
T	T	T	T
T	F	F	T
F	T	T	F
F	F	T	T

FIG. 1.7
Truth table to show that a conditional and its converse need not have the same truth value.

Having the concept of the converse of a conditional enables us to define the logical connective *if and only if* as in Definition 1.12.

Definition 1.12   For all statements p and q, the *biconditional statement* (or *biconditional*) of p with q, denoted by *p if and only if q*, is the conjunction of *if p, then q* and *if q, then p*.

From Definition 1.12 and the truth tables for *and* and *if · · · then* we can determine the truth table for *if and only if* as shown in Fig. 1.8.   From the truth table for *if and only if* we can see that:

For all statements p and q, *p if and only if q* is true whenever p and q have the same truth value and is false whenever p and q have opposite truth values.

p	q	*if p, then q*	*if q, then p*	*p if and only if q* (*if p, then q*) and (*if q, then p*)
T	T	T	T	T
T	F	F	T	F
F	T	T	F	F
F	F	T	T	T

FIG. 1.8
Truth table for if and only if.

Exercise set 1.5

1. For each of the following pairs of statements, with truth values as indicated:

    a. Write the conditional using the first statement as hypothesis and the second as conclusion, and indicate its truth value.
    b. Write the converse of each conditional in part a and indicate its truth value.
    c. Write the biconditional of each pair of statements and indicate its truth value.

        (1) London is a city. (*T*)
             Venus is a planet. (*T*)
        (2) Apples grow on trees. (*T*)
             Good apples are mushy. (*F*)
        (3) Elephants are small. (*F*)
             Monkeys are the predecessors of man. (*T*)
        (4) Potatoes are fruit. (*F*)
             Apples are not fruit. (*F*)

2. Write each of the following conditionals in the *if* · · · *then* form.

    a. A number is divisible by 2 if it is divisible by 4.
    b. John will drive only if Bill will watch the map.
    c. A statement is false whenever the negation of the statement is true.

1.5.4 Some special statement patterns

When symbols (such as $p, q, r, \ldots$) for statements are connected by one or more of the logical connectives, the combination is called a *statement pattern* provided they are connected in such a way that when symbols are replaced by statements the result is a statement. Examples of statement patterns are "*p and (not q),*" "*if p, then q,*" "*if (p or r), then q,*" . . . ; whereas, "*p and and q or*" is not a statement pattern. Some statement patterns have truth value *T* for all possible truth-value assignments to the statement symbols; such a pattern is called a *tautology*. A *contradiction* is a statement pattern which is false for all possible truth value assignments to the statement symbols in the pattern.

We illustrate by a truth table in Fig. 1.9 that

p and (not p) is a contradiction

and that

p or (not p) is a tautology.

FIG. 1.9
*Truth table to
show that* p *and*
(not p) *is a contra-
diction and* p *or*
(not p) *is a
tautology.*

p	*not p*	*p and (not p)*	*p or (not p)*
T	F	F	T
F	T	F	T

Any statement which is formed by substituting statements for statement symbols in a statement pattern which is a contradiction is also called a *contradiction*. For example, if "p" in the contradiction "*p and (not p)*" is replaced by

$4 + 5 = 9,$

the resulting statement

$4 + 5 = 9$ and $4 + 5 \neq 9$

is called a *contradiction* and is a false statement. If "p" in the tautology "*p or (not p)*" is replaced by

$4 + 5 = 9,$

the resulting statement

$4 + 5 = 9$ or $4 + 5 \neq 9$

is a true statement. (Note that the word "contradiction" is used in reference to both statement patterns and statements; whereas, the word "tautology" is used in reference to statement patterns only.)

1.6 OPERATIONS ON SETS

In this section we will define and illustrate several important operations on sets.

Definition 1.13 For all sets A and B, the *union* of A and B, denoted by $A \cup B$ (read "A union B"), is the set of elements which are in A or in B. Symbolically, for all sets A and B,

$A \cup B = \{x | x \in A \text{ or } x \in B\}$;

that is, for every x, $x \in A \cup B$ if and only if $x \in A$ or $x \in B$.

Definition 1.14   For all sets A and B, the *intersection* of A and B, denoted by $A \cap B$ (read "A intersect B"), is the set of elements which are in A and in B. Symbolically, for all sets A and B, $A \cap B = \{x | x \in A \text{ and } x \in B\}$; that is, for every x, $x \in A \cap B$ if and only if $x \in A$ and $x \in B$.

Definition 1.15   For all sets A and B, A and B are called *disjoint* if and only if $A \cap B = \emptyset$.

Definition 1.16   For every set A, the *complement* of A, denoted by $\sim A$ (read "complement of A"), is the set of elements in the universal set U, not in A. Symbolically, for every set A,

$\sim A = \{x | x \in U \text{ and } x \notin A\}$;

that is, for every x, $x \in (\sim A)$ if and only if $x \in U$ and $x \notin A$.

Definition 1.17   For all sets A and B, the *relative complement* of B with respect to A, denoted by $A \setminus B$ (read "A not B"), is the set of elements contained in A which are not in B. Symbolically, for all sets A and B, $A \setminus B = \{x | x \in A \text{ and } x \notin B\}$; that is, for every x, $x \in (A \setminus B)$ if and only if $x \in A$ and $x \notin B$.

The four preceding definitions are illustrated in Example 1.15.

EXAMPLE 1.15   Let $U = \{a,b,c,d,e,f,h,m,o,p\}$, $A = \{a,b,c,d,e\}$, $B = \{a,b,c\}$, $C = \{b,d,f,h\}$, and $D = \{m,o,p\}$.

a. $A \cup B = \{a,b,c,d,e\}$.
b. $A \cup C = \{a,b,c,d,e,f,h\}$.
c. $A \cap B = \{a,b,c\}$.
d. $A \cap C = \{b,d\}$.
e. $A \cap D = \emptyset$; that is, A and D are disjoint.
f. $A \setminus B = \{d,e\}$.
g. $B \setminus A = \emptyset$.
h. $A \setminus C = \{a,c,e\}$.
i. $C \setminus A = \{f,h\}$.
j. $\sim A = \{f,h,m,o,p\}$.
k. $\sim \emptyset = U$.
l. $\sim U = \emptyset$.

Before considering the last set operation, we need the concept of ordered pair. An *ordered pair* consists of two objects (a pair) such that one is considered to be the first (and is called the *first component*) and the other is considered to be the second (and is called the *second component*). To write the name for an ordered pair, we write the names for the first and second components in that order, separate them with a comma, and enclose them in parentheses. For all ordered pairs (a,b) and (c,d), $(a,b) = (c,d)$ if and only if $a = c$ and $b = d$.

Definition 1.18 For all sets A and B, the *cross* (or *cartesian*) *product* of A and B, denoted by $A \times B$ (read "A cross B"), is the set of all ordered pairs with elements of A as first component and elements of B as second component. Symbolically, for all sets A and B, $A \times B = \{(x,y)|x \in A \text{ and } y \in B\}$; that is, for every (x,y), $(x,y) \in A \times B$ if and only if $x \in A$ and $y \in B$.

An illustration of the operation of cross product is given in Example 1.16.

EXAMPLE 1.16 Let $P = \{1,2\}$ and $Q = \{a,b,c\}$.

a. $P \times Q = \{(1,a),(1,b),(1,c),(2,a),(2,b),(2,c)\}$.
b. $Q \times P = \{(a,1),(a,2),(b,1),(b,2),(c,1),(c,2)\}$.

Some additional illustrations of the set operations are given in Examples 1.17 and 1.18.

EXAMPLE 1.17 Let $P = \{x|x$ is a counting number and $x > 2\}$ and $Q = \{x|x$ is a counting number and $x < 27\}$.

a. $P \cup Q = \{x|x$ is a counting number, and $x > 2$ or $x < 27\}$.
b. $P \cap Q = \{x|x$ is a counting number, $x > 2$, and $x < 27\}$.
c. $P \times Q = \{(x,y)|x$ and y are counting numbers, $x > 2$, and $y < 27\}$.
d. $P \setminus Q = \{x|x$ is a counting number, $x > 2$, and $x \not< 27\}$.

The following are true statements.

$5 \in P \cup Q.$ $900 \in P \cup Q.$
$2 \notin P \cap Q.$ $5 \in P \cap Q.$
$(4,19) \in P \times Q.$ $(5,29) \notin P \times Q.$
$28 \in P \setminus Q.$ $23 \notin P \setminus Q.$

EXAMPLE 1.18 It is important to be familiar with statements such as the following:

a. $5 \in A \cup B$, $\therefore 5 \in A$ or $5 \in B$.
b. For every x, if $x \in X \cup Y$, then $x \in X$ or $x \in Y$.
c. $9 \in A \cap B$, $\therefore 9 \in A$ and $9 \in B$.
d. For every x, if $x \in X \cap Y$, then $x \in X$ and $x \in Y$.
e. $(2,3) \in C \times D$, $\therefore 2 \in C$ and $3 \in D$.
f. For every (x,y), if $(x,y) \in X \times Y$, then $x \in X$ and $y \in Y$.
g. $5 \in A \setminus B$, $\therefore 5 \in A$ and $5 \notin B$.
h. For every x, if $x \in X \setminus Y$, then $x \in X$ and $x \notin Y$.

We call the reader's attention to some special properties that the empty set enjoys with respect to the operations of union and intersection. Namely:

1. For every set A, $A \cup \emptyset = A$.
2. For every set A, $A \cap \emptyset = \emptyset$.

Some special properties of sets which we will find useful in our later study of geometry are the following:

1. For all sets A and B, $A \cap B \subseteq A$ and $A \cap B \subseteq B$.
2. For all sets A and B, if $A \subseteq B$, then $A \cap B = A$.
3. For all sets A and B, if $A \subseteq B$, then $A \cup B = B$.
4. For all sets A, B, and C, if $A \subseteq B$ or $A \subseteq C$, then $A \subseteq B \cup C$.
5. For all sets A, B, and C, if $A \subseteq B$ and $A \subseteq C$, then $A \subseteq B \cap C$.
6. For all sets A, B, and C, if $A \subseteq B$ and $B \subseteq C$, then $A \subseteq C$.

Exercise set 1.6

1. Let $A = \{5,10,15\}$, $B = \{10,15,20\}$, $D = \{10,15\}$, $C = \{1,2,3, \ldots\}$, $C_e = \{2,4,6, \ldots\}$, $X = \{x | x \in C \text{ and } x > 5\}$, and $Y = \{x | x \in C \text{ and } x < 16\}$.

   a. Write the following sets:

(1) $A \cup B$	(2) $A \cap B$	(3) $A \cap D$
(4) $X \cup Y$	(5) $A \cup Y$	(6) $A \cap Y$
(7) $X \cap Y$	(8) $A \setminus C$	(9) $C \setminus C_e$
(10) $A \times B$	(11) $D \times D$	(12) $X \times Y$
(13) $B \setminus D$	(14) $X \setminus Y$	(15) $Y \setminus A$

    b. Indicate whether the following are true or false:

        (1) $(10,10) \in B \times D$.   (2) $(5,4) \in A \times X$.
        (3) $(12,19) \in Y \times X$.   (4) $(12,19) \in X \times Y$.
        (5) $(2,5) \in C \times C_e$.   (6) $(5,2) \in C_e \times C$.
        (7) $5 \in A \cup B$.   (8) $(5,10) \in A \cup B$.
        (9) $6 \in C \cap X$.   (10) $9 \in X \cap Y$.
        (11) $5 \in A \setminus C$.   (12) $3 \in A \setminus C$.
        (13) $6 \in X \setminus Y$.   (14) $21 \in X \setminus Y$.
        (15) $\{21\} \in X \setminus Y$.

2. Let $Z = \{a,b,c\}$, and $Q = \{d,e\}$. Write the following sets:

    a. $Z \cup \varnothing$ **b.** $Q \cup \varnothing$ **c.** $Z \cap \varnothing$
    d. $Q \cap \varnothing$ **e.** $Z \times \varnothing$ **f.** $Q \times \varnothing$

3. Let $U = \{1,2,3,4,5,6,7,8,9,10\}$, $A = \{2,3,4,5,6\}$, $B = \{4,5,6,7\}$, $C = \{x | x \in U$ and $x < 2\}$, and $D = \varnothing$. Write the following sets:

    a. $\sim A$ **b.** $\sim C$
    c. $\sim D$ **d.** $\sim U$
    e. $\sim \varnothing$ **f.** $\sim (\sim A)$
    g. $A \cap (\sim A)$ **h.** $A \cup (\sim A)$
    i. $\sim (A \cap B)$ **j.** $(\sim A) \cup (\sim B)$
    k. $\sim (A \cup B)$ **l.** $(\sim A) \cap (\sim B)$
    m. $\sim (U \cap D)$ **n.** $(\sim U) \cap (\sim D)$
    o. $(\sim U) \cup (\sim D)$

4. Use the definitions of the set operations to make true statements of the following (*hint:* see Example 1.18):

    a. $1 \in B \cup C, \therefore 1 \in$ _____ or $1 \in$ _____.
    b. $2 \in X$ and $2 \in Y, \therefore 2 \in$ _____.
    c. $6 \in A \cap B, \therefore$ _____.
    d. $(2,3) \in A \times Y, \therefore$ _____.
    e. $(9,9) \in M \times N, \therefore$ _____.
    f. $6 \in R$ and $5 \in S, \therefore (6,5) \in$ _____.
    g. $12 \in R$ and $12 \in S, \therefore 12 \in$ _____.
    h. $12 \in R$ and $12 \in S, \therefore (12,12) \in$ _____.
    i. $5 \in B$ or $5 \in C, \therefore$ _____.

1.7 RELATIONS

Certain concepts which are important in mathematics are already familiar to us from everyday situations, and in these cases it remains only to formalize the concepts; that is, to define them precisely. *Relation* is such a concept; we will explore it by abstracting from the "everyday" notion of relation. We often make reference to our "relations," meaning spouses, cousins, uncles, and so on. We will explore one of these relations and abstract from it to arrive at the precise mathematical concept.

Consider the relation "is the brother of." First of all, we note that when we refer to this relation we have in mind some people; that is a set! It may be the set of all people or some subset of it. In an investigation of this relation it is meaningful to ask whether certain statements are true or false; to wit:

1. Bill is the brother of Jane.
2. Cindy is the brother of Bob.
3. Joe is the brother of James.
4. Jack is the brother of Ron.

$\vdots$

If statement 1 is true, then we have identified or singled out a *pair* of people. Moreover, for statement 1 to be true, the order in which "Bill" and "Jane" appear in the statement is important. (The statement "Jane is the brother of Bill" is false.) Therefore, we have identified or singled out an *ordered pair*. If, on the other hand, statement 3 is false, then we exclude the ordered pair (Joe, James). We see that the phrase "is the brother of" sorts out or defines a *set of ordered pairs*. Mathematically, we would say that the ordered pair (Bill, Jane) is an element of the relation "is the brother of." We would say that the ordered pairs (Jane, Bill) and (Joe, James) do not belong to the relation. These considerations bring us very close to the precise concept of a *relation on a set*.

Definition 1.19 A *relation R on a set S* is a subset of $S \times S$; that is, a relation is a set of ordered pairs. The set of first components is called the *domain* of the relation. The set of second components is called the *range*.

If R is a relation on S and x and y are elements of S such that x is related to y by R, that is, $(x,y) \in R$, we write $x \, R \, y$. We illustrate further with our example. Let S be the set of all people in the United States at some given instant. Let R be the relation on S, defined by "is the brother of." Then R is the set of ordered pairs (P_1,P_2) of people (subset of $S \times S$) for which it is true that

P_1 is the brother of P_2. To indicate that P_1 is related to P_2 by "is the brother of," we usually write

P_1 is the brother of P_2

or

$P_1 \, R \, P_2$.

To write $(P_1,P_2) \in R$ means the same as $P_1 \, R \, P_2$. In set notation we describe the relation R as

$R = \{(P_1,P_2) \in S \times S | P_1 \text{ is the brother of } P_2\}$.

We have, in addition to the concept of a relation on a set, the concept of a relation *between* two sets which is given in Definition 1.20.

Definition 1.20 A *relation R between sets A and B* (or a *relation from A to B*) is a subset of $A \times B$; that is, a relation is a set of ordered pairs. The set of first components is called the *domain* of the relation. The set of second components is called the *range*.

Note: If $A = B$, then Definition 1.20 defines the concept of a relation on A.

The concepts of a relation on a set and a relation between sets are illustrated in Example 1.19.

EXAMPLE 1.19 Let $A = \{1,2,3,4\}$ and $B = \{a,b,c\}$. Then,

$R = \{(1,2),(3,4),(1,1)\}$ is a relation on A.
$Q = \{(a,b)\}$ is a relation on B.
$M = \{(a,a),(b,c)\}$ is a relation on B.
$N = \{(1,a),(3,a),(1,b)\}$ is a relation from A to B.
$O = \{(b,3),(a,2)\}$ is a relation from B to A.

It is frequently convenient to use set builder notation in order to define a relation as illustrated in Example 1.20.

EXAMPLE 1.20 Let $A = \{1,2,3, \ldots ,20\}$ and $B = \{2,4,6, \ldots , 26\}$. Then $R = \{(x,y) \in A \times B | y = x + 1\}$ is a relation from A to B.

The following are true statements about the relation R:

$(1,2) \in R.$ $(25,26) \notin R.$
$(2,3) \notin R.$ $(21,22) \notin R.$
$(5,6) \in R.$

In this case, elements $(x,y) \in R$ are related by the open sentence $y = x + 1$.

EXAMPLE 1.21 Let $A = \{1\}$, $B = \{a,b\}$, and $C = \{e,f\}$. Every subset of $(A \times B) \times C$ is a relation from $A \times B$ to C. That is, every subset of $(A \times B) \times C$ is a relation whose domain is a subset of $A \times B$ and range is a subset of C. Thus $(A \times B) \times C$ contains ordered pairs whose first components are elements of $A \times B$ (that is, also ordered pairs) and second components are elements of C. That is,

$A \times B = \{(1,a),(1,b)\}$

so that

$(A \times B) \times C = \{((1,a),e),((1,a),f),((1,b),e),((1,b),f)\}.$

The following are all examples of relations from $A \times B$ to C:

$R_1 = \{((1,a),e)\}.$
$R_2 = \{((1,a),f),((1,b),f)\}.$
$R_3 = \{((1,b),f)\}.$
$R_4 = (A \times B) \times C.$

There are still more relations from $A \times B$ to C. How many?

We sometimes think of a relation as being a *rule* which determines a pairing between elements of one set, and elements of another (or the same) set. Of course, a pairing of elements according to some rule does indeed define a set of ordered pairs. A relation is sometimes said to *map* elements of its domain onto elements of its range. Relations can be classified according to the "number" of elements of the domain which get mapped onto one or more elements of the range. In this sense we talk about relations which are *one-to-one, one-to-many, many-to-one,* or *many-to-many*. Frequently a *mapping diagram* is helpful in visualizing a relation. We use mapping diagrams to help illustrate the above classification of relations.

Figure 1.10 pictures two relations: $R_1 = \{(1,2),(3,4),(5,6),(7,8)\}$ and $R_2 = \{(1,2),(3,8),(7,4)\}$ from $S = \{1,3,5,7\}$ to $T = \{2,4,6,8\}$. Note that R_1 is a *one-to-one correspondence* while R_2 is a *one-to-one relation* but is *not* a one-to-one correspondence. Note that every one-to-one correspondence is a one-to-one relation, but not every one-to-one relation is a one-to-one correspondence. To be a *one-to-one correspondence* between two sets S and T *every* element of S must be the first component of exactly one pair and *every* element of T must be the second component of exactly one pair.

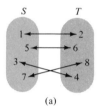
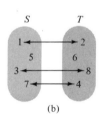

(a)   (b)

FIG. 1.10
(a) *One-to-one correspondence.* (b) *One-to-one relation which is not a one-to-one correspondence.*

The relation $R_3 = \{(1,2),(1,3),(1,4),(2,5),(2,6),(3,7)\}$, which is diagrammed in Fig. 1.11, is a one-to-many relation from $X =$

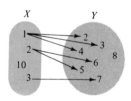

FIG. 1.11
One-to-many relation.

$\{1,2,3,10\}$ to $Y = \{2,3,4,5,6,7,8\}$. In Fig. 1.12, the many-to-one relation $R_4 = \{(1,b),(2,b),(3,b),(5,a),(6,a),(7,5)\}$ from set Z to set W is diagrammed, where $Z = \{1,2,3,4,5,6,7\}$ and $W = \{a,b,5\}$.

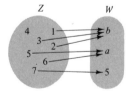

FIG. 1.12
Many-to-one relation.

After these illustrations the reader can write and diagram examples of many-to-many relations from $Q = \{a,b,c,d\}$ to $P = \{1,2,3,4,5\}$; in doing so you will discover that a many-to-many relation is both many-to-one and one-to-many.

Exercise set 1.7

1. Let $S = \{0,1,2,3,4\}$.

   a. Let R be the relation defined on S as follows:

   For all $x, y \in S$, $x \, R \, y$ if and only if $x < y$.

   Write R as a set of ordered pairs.
   b. Let Q be the following relation on S:

   $$Q = \{(x,y) \in S \times S \, | \, x \cdot y = 0\}.$$

   Write all the ordered pairs in Q.
   c. Let T be the relation on S defined as follows:

   For all $x, y \in S$, $x \, T \, y$ if and only if $x + y = 5$.

   Write T as a set of ordered pairs.
   d. Let M be the relation defined on S as follows:

   For all $x, y \in S$, $x \, M \, y$ if and only if $x - y = 2$.

   Write M as a set of ordered pairs.

2. Let S be the set of all people in the United States. Let R be a relation on S defined as "is the husband of."

   a. Write some ordered pairs that you know to be elements of this relation.
   b. Is it possible (physically and legally) that a person A stand in this relation to more than one person? That is, can a person be the first component of more than one ordered pair in R?
   c. Answer questions a and b above for the following relations on S:

   (1) is the mother of   (2) is the friend of
   (3) is older than   (4) is the daughter of

3. Let $A = \{a,b\}$ and $B = \{c,d,e\}$.

   a. $A \times B = ?$.   b. $A \times A = ?$.
   c. $B \times B = ?$.   d. $B \times A = ?$.
   e. Write at least three relations from A to B.
   f. Write at least three relations from B to A.
   g. Write at least three relations on A.
   h. Write at least three relations on B.

4. Let $X = \{1,2,3\}$ and $Y = \{9,10,11\}$.

$R_1 = \{(1,9),(1,10),(1,11)\}$,
$R_2 = \{(1,9),(2,9)\}$,
$R_3 = \{(1,9),(2,10),(3,11)\}$,

are all relations from X to Y. What is the domain and range, respectively, of R_1, R_2, R_3?

5. Let $C = \{1,2,3,4, \ldots\}$. Use an open sentence and set builder notation to define the following relations on C.

    a. $\{(1,2),(2,3),(3,4), \ldots\}$
    b. $\{(1,1),(2,4),(3,9), \ldots\}$
    c. $\{(2,5),(4,7),(6,9), \ldots\}$
    d. $\{(2,1),(3,2), \ldots\}$
    e. $\{(1,8),(8,1),(2,7),(7,2),(6,3),(3,6),(4,5),(5,4)\}$

6. Let $D = \{0,1,2,3,4,5\}$ and $E = \{1,2,3,4,5,6,7,8\}$. Write the following relations by listing the ordered pairs.

    a. $R_1 = \{(x,y) \in D \times E \mid y = x\}$.
    b. $R_2 = \{(x,y) \in D \times E \mid y = x + 3\}$.
    c. $R_3 = \{(x,y) \in D \times E \mid y = 2x\}$.
    d. $R_4 = \{(x,y) \in D \times E \mid y = x \div 2\}$.

7. Let $A = \{a,b\}$, $B = \{c,d\}$, and $C = \{1\}$.

    a. Write the set $A \times B$.
    b. Write the set $(A \times B) \times C$.
    c. Write a relation from $A \times B$ to C.
    d. Write $B \times C$.
    e. Write $A \times (B \times C)$.
    f. Write a relation from A to $B \times C$.

8. Let $A = \{a,b,c,d,e\}$ and $B = \{1,2,3,4,5,6,7,8,9\}$. Draw a mapping diagram of each of the following relations and indicate which of the following apply: one-to-one correspondence, one-to-one, one-to-many, many-to-one, many-to-many:

    a. $R_1 = \{(a,1),(a,2)\}$.
    b. $R_2 = \{(a,1),(b,6),(c,7),(d,8),(d,9)\}$.
    c. $R_3 = \{(a,1),(b,1),(c,1),(d,2)\}$.
    d. $R_4 = \{(a,1),(b,1),(c,2),(d,2)\}$.
    e. $R_5 = \{(a,5),(a,6),(a,7)\}$.

f. $R_6 = \{(a,6),(b,8),(c,8)\}.$

g. $R_7 = \{(a,6),(b,7),(c,8),(d,9),(e,1)\}.$

h. $R_8 = \{(a,5),(c,5),(c,6),(c,7),(c,8)\}.$

1.8 REFLEXIVE, SYMMETRIC, TRANSITIVE, AND EQUIVALENCE RELATIONS

To illustrate three special properties which some relations enjoy, we will take a familiar relation and abstract from it the general idea we are after.

Let S denote the set of all people in the United States, and let H be the relation defined on S as "is the same height as" (correct to the nearest inch). We make the trivial observation that

For every $P \in S, P\,H\,P$

or

For every $P \in S, (P,P) \in H.$

That is, H is a relation which satisfies the *reflexive* property or is a *reflexive relation*. We also observe that

For all $P, Q \in S$, if $P\,H\,Q$, then $Q\,H\,P$

or

For all $P, Q \in S$, if $(P,Q) \in H$, then $(Q,P) \in H.$

That is, H is a relation which satisfies the *symmetric* property or is a *symmetric relation*. We observe finally that

For all $P, Q, R \in S$, if $P\,H\,Q$ and $Q\,H\,R$, then $P\,H\,R$

or

For all $P, Q, R \in S$, if $(P,Q) \in H$ and $(Q,R) \in H$, then $(P,R) \in H.$

That is, H is a relation which satisfies the *transitive* property or is a *transitive relation*.

Definition 1.21 Let S be a nonempty set and R a relation on S:

a. R is a *reflexive relation* on S (or has the *reflexive* property) if and only if

for every $x \in S, x\,R\,x.$

b. *R* is a *symmetric relation* on *S* (or has the *symmetric* property) if and only if

for all $x, y \in S$, if $x R y$, then $y R x$.

c. *R* is a *transitive relation* on *S* (or has the *transitive* property) if and only if

for all $x, y, z \in S$, if $x R y$ and $y R z$, then $x R z$.

A relation which enjoys all three of the properties—reflexive, symmetric, and transitive—is of such importance in mathematics that we give it the special name of *equivalence relation*.

Definition 1.22 An *equivalence relation E* on a set *S* is a relation on *S* which satisfies:

1. *The Reflexive Property.* For every $x \in S$, $x E x$.
2. *The Symmetric Property.* For all $x, y \in S$, if $x E y$, then $y E x$.
3. *The Transitive Property.* For all $x, y, z \in S$, if $x E y$ and $y E z$, then $x E z$.

The three properties of an equivalence relation are further illustrated in Example 1.22.

EXAMPLE 1.22 Let

$P = \{20, 40, 60, 120, 960, 260, 940, 440, 140, 560, 320, 420\}$.

Let $\wedge$ be the relation defined on *P* as follows: For all $x, y \in P$, $x \wedge y$ if and only if the numeral for *x* has the same tens digit as the numeral for *y*. (The tens digit of 60 is 6, of 940 is 4.) We leave to the reader the problem of showing that $\wedge$ enjoys the reflexive, symmetric, and transitive properties and is therefore an equivalence relation on *P*.

Exercise set 1.8

1. Let *S* be the set of people in the United States. Determine whether the following relations defined on *S* have:

    (a) the reflexive property,
    (b) the symmetric property,
    (c) the transitive property.

a. is the brother of   b. has the same parents as
c. is the same age as   d. is older than

2. We called "subset," "proper subset," and "one-to-one correspondence" relations on sets. For each of these relations indicate whether it has the reflexive, symmetric, and transitive
properties.

3. Let $Q = \{1,2,3,4,5,41,111,403,22,42,301,44,444,15732,974,16234\}$.
Let : be the relation defined on Q as follows:

For every $x, y \in Q$, $x : y$ if and only if the numeral for x has
the same ones digit as the numeral for y.

a. Write : as a set of ordered pairs.
b. Show that : is an equivalence relation.
c. Indicate whether the following are true or false:

    (1) $1 : 41$.   (2) $4 : 15732$.
    (3) $403 : 3$.   (4) $22 : 4$.

4. Let

$$P = \{200,111,112,104,100,101,300,21,40,311,401,1000,1001,221,$$
$$500,102\}.$$

Let $\sim$ be the relation defined on P as follows:

For all $x, y \in P$, $x \sim y$ if and only if "the sum of the digits of x
equals the sum of the digits of y."

(*Note:* The sum of the digits of 221 is $2 + 2 + 1 = 5$.)

a. Write $\sim$ as a set of ordered pairs.
b. Show that $\sim$ is an equivalence relation on P.

5. Let $C = \{1,2,3,4, \ldots\}$. Let $\equiv$ be defined on C as follows:

For all $x, y \in C$, $x \equiv y$ if and only if x and y have the same re-
mainder when divided by 4.

a. Show that $\equiv$ is an equivalence relation on C.
b. Indicate whether the following are true or false:

    (1) $1 \equiv 5$.   (2) $2 \equiv 17$.
    (3) $18 \equiv 22$.   (4) $7 \equiv 14$.
    (5) $1763 \equiv 672$.

1.9 FUNCTIONS

By making certain restrictions on the general notion of relation, we arrived at the idea of an equivalence relation. By making a different restriction on this general idea, we arrive at the concept of a function. A *function from A to B* is a one-to-one, or many-to-one relation such that *every element of A* has exactly one member of *B* paired with it. Figures 1.13–1.16 illustrate relations which are and are not functions from *P* to *Q*.

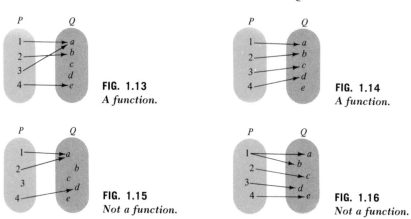

FIG. 1.13
A function.

FIG. 1.14
A function.

FIG. 1.15
Not a function.

FIG. 1.16
Not a function.

The formal definition of function is given in Definition 1.23.

Definition 1.23 A *function F from A to B* is a relation from *A* to *B* such that:

1. For every $x \in A$ there exists $y \in B$ such that $(x,y) \in F$.
2. For every $x \in A$, if $(x,y) \in F$ and $(x,z) \in F$, then $y = z$.

Statement 1 says simply that every element of *A* must be a first component of an ordered pair; 2 says that no element of *A* can be the first component of more than one pair. Statement 2 can also be stated as follows:

If $(x,y) \in F$ and $(z,w) \in F$ such that $x = z$, then $y = w$.

Property 2 in the definition of function is sometimes called the *well-defined* property. We say that a relation which has this property is a well-defined function (from *A* to *B*). We will see this property again in Section 9.1, in connection with operations. If $A = B$ in Definition 1.23, we call F a function on A.

The concept of a function is further illustrated in Examples 1.23–1.25.

EXAMPLE 1.23 Let $X = \{a,b,c\}$ and $Y = \{d,e,f,g\}$.

$F_1 = \{(a,d),(b,f),(c,f)\}$ is many-to-one and is a function from X to Y.

$F_2 = \{(a,c),(a,d),(b,e),(c,g)\}$ is not a function from X to Y because a is the first component of more than one pair.

$F_3 = \{(a,d),(b,d),(c,d)\}$ is many-to-one and is a function from X to Y.

$F_4 = \{(a,d),(b,g)\}$ is not a function from X to Y since c is not the first component of any pair.

EXAMPLE 1.24 Let $C = \{1,2,3,4, \ldots\}$ and $C_e = \{2,4,6, \ldots\}$.

$F_1 = \{(x,y) \in C \times C | y = x + 1\}$ is a function from C to C, or a function on C.

$F_2 = \{(x,y) \in C \times C_e | y = 2x\}$ is a function from C to C_e.

$F_3 = \{(x,y) \in C \times C_e | y = x - 9\}$ is not a function from C to C_e since $1 - 9, 2 - 9, 3 - 9, \ldots, 9 - 9$ are not elements of C_e; therefore, 1, 2, 3, $\ldots$, 9 cannot be the first component of any pair in F_3.

EXAMPLE 1.25 Let $S = \{1,2\}$. Then

$S \times S = \{(1,1),(1,2),(2,1),(2,2)\}$.

$F = \{((1,1),1),((1,2),1),((2,1),2),((2,2),2)\}$

is a function from $S \times S$ to S.

If F is a function from A to B (or from A to A), then: for every $x \in A$, the element that is paired with x is called the *image of x* (or for emphasis, the *F-image of x*, or the *image of x under F*), and is denoted by $F(x)$. The notation $F(x)$ is read "the F-image of x," or "the value of F at x," or simply as "F of x." If F is a function from A to B, we frequently indicate this by writing $F: A \to B$; moreover, if $x \in A$ and y is the F-image of x, we will indicate this by writing $F(x) = y$, $F: x \to y$, or $x \xrightarrow{F} y$, where $F(x) = y$ is usually read "the value of F at x is y," or "F of x equals y." $F{:}x \to y$ and $x \overset{F}{\to} y$ are usually read as "x gets mapped by F onto y," or "the F-image of x is y."   It is important to take note of the fact that if F is a function from A to B, then for every $x \in A$, $F(x) \in B$.

We illustrate some of these ideas in Example 1.26.

EXAMPLE 1.26

a.   Let $P = \{1,2,3\}$ and $Q = \{3,4,5,6\}$.   Let f be the function from P to Q defined by $f = \{(1,3),(2,4),(3,5)\}$.
Then, the f-image of 1 is 3, that is, $f(1) = 3$, $f{:}1 \to 3$, $1 \overset{f}{\to} 3$.

The f-image of 2 is 4, that is, $f(2) = 4$, $f{:}2 \to 4$, $2 \overset{f}{\to} 4$.

The f-image of 3 is 5, that is, $f(3) = 5$, $f{:}3 \to 5$, $3 \overset{f}{\to} 5$.

b.   Let $S = \{2,3\}$; then $S \times S = \{(2,2),(2,3),(3,2),(3,3)\}$.   Let g be the function from $S \times S$ to S such that: $(2,2) \overset{g}{\to} 2$, $(2,3) \overset{g}{\to} 3$, $(3,2) \overset{g}{\to} 2$, and $(3,3) \overset{g}{\to} 3$.   Then g is the function defined by the following set of ordered pairs:

$$g = \{((2,2),2),((2,3),3),((3,2),2),((3,3),3)\}.$$

c.   Let g be the function from $M = \{a,b,c,d\}$ to $N = \{e,f\}$ defined by the accompanying diagram.   We see that:

(1)   The g-image of a is e; that is, $g(a) = e$, $g{:}a \to e$, $a \overset{g}{\to} e$.

(2)   The g-image of b is e; that is, $g(b) = e$, $g{:}b \to e$, $b \overset{g}{\to} e$.

(3)   The g-image of c is f; that is, $g(c) = f$, $g{:}c \to f$, $c \overset{g}{\to} f$.

(4)   The g-image of d is f; that is, $g(d) = f$, $g{:}d \to f$, $d \overset{g}{\to} f$.

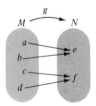

We mentioned in the previous section that a function from A to B can be thought of as a rule which assigns to every $x \in A$, exactly one element of B.   The notation we have just introduced makes it natural to exploit this idea more fully.   For example, we can define a function f from $C = \{1,2,3,4, \ldots\}$ to C as follows: For every $x \in C$, $f(x) = x + 1$. Many elementary school textbooks refer to "$x + 1$" as the *function rule* and "$f(x) = x + 1$" as the *function equation*. The equation $f(x) = x + 1$ gives us a way to determine the image $f(x)$ for every $x \in A$. Some illustrations are: $f(1) = 1 + 1 = 2$, $f(2) = 2 + 1 = 3$, $f(3) = 3 + 1 = 4$, ...

The three notational schemes we use in defining functions are illustrated and compared in Example 1.27.

EXAMPLE 1.27 Let $A = \{1,2,3,4\}$ and $B = \{1,2,3,4,5,6,7\}$. Let g be the function from A to B such that g of x equals $2x - 1$ for every x in A. "Shorthand" methods to write this are

$g : A \to B$ such that for every $x \in A$, $g(x) = 2x - 1$,
$g : A \to B$ such that for every $x \in A$, $g : x \to 2x - 1$,
$g : A \to B$ such that for every $x \in A$, $x \xrightarrow{g} 2x - 1$.

In set builder notation g can be defined by

$g = \{(x, g(x)) \in A \times B \mid g(x) = 2x - 1\}$

or

$g = \{(x, y) \in A \times B \mid y = 2x - 1\}$.

In any case we have the following information about g: The domain of g is the set A; that is, the only permissible replacements for x in the open sentence $g(x) = 2x - 1$ are elements of A.

$g(1) = 2 \cdot 1 - 1 = 1$; that is, $(1,1) \in g$.
$g(2) = 2 \cdot 2 - 1 = 3$; that is, $(2,3) \in g$.
$g(3) = 2 \cdot 3 - 1 = 5$; that is, $(3,5) \in g$.
$g(4) = 2 \cdot 4 - 1 = 7$; that is, $(4,7) \in g$.

Therefore, $g = \{(1,1), (2,3), (3,5), (4,7)\}$.
The range R_g of g is: $R_g = \{1,3,5,7\}$. (*Note:* $R_g \subseteq B$.)

Exercise set 1.9

1. Let $A = \{a,b,c,d\}$ and $B = \{e,f,g,h\}$. For the sets given below indicate which of the following apply:

one-to-one correspondence   many-to-many
one-to-one   relation from A to B
many-to-one   function from A to B
one-to-many

    a. $\{(a,e),(b,f),(c,g),(d,h)\}$
    b. $\{(a,e),(b,e),(c,f),(d,f)\}$
    c. $\{(a,e),(b,e),(c,e)\}$
    d. $\{(a,f),(b,f),(c,f),(d,f)\}$
    e. $\{(a,e),(a,f),(a,g),(a,h)\}$
    f. $\{(a,e),(a,f),(b,e),(c,e),(d,g)\}$
    g. $\{(e,a),(f,b),(g,c),(h,d)\}$
    h. $\{(a,h),(b,g),(c,f)\}$
    i. $\varnothing$
    j. $\{(a,e)\}$

2. Draw a mapping diagram of each of the relations in Exercise 1.

3. Let $A = \{a,b\}$ and $C = \{e,f\}$.

    a. $A \times A = ?$.
    b. Is $R = \{((a,a),e),((a,b),e),((b,b),f),((b,a),f)\}$ a function from $A \times A$ to C?
    c. Write another function from $A \times A$ to C.

4. Let $S = \{0,1,2\}$.

    a. $S \times S = ?$.
    b. Is $R = \{((0,0),0),((0,1),1),((0,2),2),((1,0),0),((1,1),1),$ $((1,2),2),((2,0),0),((2,1),1),((2,2),2)\}$ a function from $S \times S$ to S?
    c. Write another function from $S \times S$ to S.

5. Let $C = \{1,2,3,4, \ldots\}$.

    a. If $R_1 = \{(x,y) \in C \times C | y = x\}$, is R_1 a function on C?
    b. If $R_2 = \{(x,y) \in C \times C | y = x + 1\}$, is R_2 a function on C?
    c. If $R_3 = \left\{(x,y) \in C \times C \left| y = \frac{x}{2} \right.\right\}$, is R_3 a function on C?
    d. If $R_4 = \{(x,y) \in C \times C | y = x - 2\}$, is R_4 a function on C?
    e. If $R_5 = \left\{(x,y) \in C \times C \left| y = \frac{x+1}{x} \right.\right\}$, is R_5 a function on C?
    f. If $R_6 = \{((1,1),2),((1,2),3),((2,1),3),((2,2),4),((1,3),4),$ $((3,1),4),((1,4),5), \ldots\}$, is R_6 a function from $C \times C$ to C?

6. Let $X = \{2,4,6,8\}$ and $Y = \{4,16,36,64\}$. Let $f = \{(2,4),(4,16),(6,36),(8,64)\}$.

    a. Is f a function from X to Y?

b. $f(2) =$ _____ , $f(4) =$ _____ , $f(6) =$ _____ , $f(8) =$ _____ .

c. Complete the following definition of f by giving the "rule" for f:

$f: X \to Y$ such that for every $x \in X$, $f(x) =$ _____ .

d. Write the function f using set builder notation; that is,

$f = \{(x, f(x)) \in$ | $\}$

or

$f = \{(x, y) \in$ | $\}$.

e. What is the domain of f?

f. What is the range of f?

7. Let $C = \{1, 2, 3, 4, \ldots\}$. Let g be the function:

$g: C \to C$ such that for every $x \in C$, $g(x) = x^2 + 2$.

a. $g(3) = (\quad)^2 + 2 =$ _____ .

b. $g(2) =$ _____ .

c. $g(4) =$ _____ .

d. $g: 7 \to (\quad)^2 + 2 =$ _____ .

e. $g: 6 \to$ _____ .

f. $7 \xrightarrow{g}$ _____ .

g. $8 \xrightarrow{g}$ _____ .

h. Indicate whether the following are true or false:

    (1) $(2,4) \in g$.   (2) $(16,4) \in g$.

    (3) $(0,0) \in g$.   (4) $(5,27) \in g$.

i. Give the missing component so that the following will be true:

    (1) $(3,\_\_) \in g$.   (2) $(7,\_\_) \in g$.   (3) $(\_\_,83) \in g$.

    (4) $(11,\_\_) \in g$.   (5) $(\_\_,171) \in g$.   (6) $(5,\_\_) \in g$.

8. The accompanying figure is a mapping diagram of a relation h from A to B.

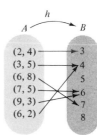

    a. Is h a function from A to B? Explain your answer.

    b. What is the domain of h? The range of h?

    c. $h((2,4)) = $ _____.

        $h:(3,5) \rightarrow$ _____.

        $(7,5) \xrightarrow{h}$ _____.

        $h((9,3)) = $ _____.

        _____ $\xrightarrow{h}$ 7.

    d. Write h as a set of ordered pairs.

    e. Describe h in words.

2

finite geometry and an introduction to euclidean geometry

2.1 INTRODUCTION

We begin the study of geometry with two broad objectives in mind. One, we wish to explore and give rather careful definition to many geometric concepts which are essential for prospective teachers. A praiseworthy endeavor, which goes well beyond the space limitations of this text, is to develop all of these geometric concepts from a minimal set of postulates. Although we will stop short of this extensive development, we shall begin in this rather formal manner. This will illustrate how such a development might progress and partially fulfill our second objective; namely, to illustrate rather carefully the role of undefined terms and postulates in a mathematical system.

Traditionally, geometry has been the subject which was considered the classic exhibit of a deductive system. A deductive system is developed from specific undefined concepts, postulates which specify how these undefined concepts are related, definitions, and theorems deducible from these.

In geometry "point" and "line" are frequently taken as undefined words. Now, pedagogically, we are probably on shaky ground with you, the reader. You are probably responding that you do in fact know what points and lines are so all this business about their being undefined concepts seems to be a bit of double talk. However, your concept of point and line is abstracted from your "real-world" experiences. If one could divorce himself from these experiences, a different notion of point and line might develop. To illustrate this and to demonstrate more clearly the role of undefined terms and postulates we digress into some mental gymnastics and develop a finite geometry.

2.2   A FINITE GEOMETRY

For our deductive development of a finite geometry we accept the following as undefined terms:

1. Point
2. Line
3. A point is on a line (or a line contains a point).

Definition DF1   The set of all points S is called *space*.

The following postulates are assumptions about the undefined terms and how they are related:

PF0   Every line is a subset of S.

PF1   Every two lines have exactly one point in common.

PF2   Every point in S is on eactly two lines.

PF3   There are exactly four lines in S.

From these postulates several theorems are deducible.

Theorem TF1   Not all lines contain the same point.

PROOF

1. There are exactly four lines, call them *l*, *m*,   1.   PF3
   n, and *p*.
2. Every two lines have exactly one point in   2.   PF1
   common; that is, the lines *l* and *m* have exactly one point in common, call it *A*. The

lines l and n have exactly one point in common, call it B.

3.	Assume $A = B$.	3.	Assumption
4.	Then, A is on l, A is on m, and A is on n.	4.	Step 2
5.	A is on exactly two lines.	5.	PF2
6.	The assumption that $A = B$ leads to the contradiction that the point A is on *more* than two lines (Step 4), and on *exactly two* lines (Step 5); therefore $A \neq B$, and the lines l, m, and n do not all contain the same point.		

We assumed, in Postulate PF3, that our system has exactly four lines. How many points are in the system? Are you sure there are any at all? Look carefully at PF3 and PF1. Because of PF3, there are *at least* two lines and by PF1 every two lines contain exactly one point. Therefore, there is at least one point! Are there more? Is the set of all points an infinite set? The question of exactly how many points are in our system can be answered by answering the question of how many *distinct* pairs of lines can be made from the four lines; call the lines l, m, n, and p and experiment: l and m form one pair, l and n another, you write the rest. We answer the question formally in a sequence of three theorems.

Theorem TF2 There are at least six points in S.

PROOF

1.	There are exactly four lines, call them l, m, n, and p.	1.	PF3
2.	l and m have exactly one point in common—A. l and n have exactly one point in common—B. l and p have exactly one point in common—C. m and n have exactly one point in common—D. m and p have exactly one point in common—E. n and p have exactly one point in common—F.	2.	PF1

3. Assume that not all of A, B, C, D, E, and F are different; that is, assume that there is less than six points. Suppose $C = B$. **3.** Assumption

4. $\therefore$ l and n have point C in common and l and p have point C in common. **4.** Step 2

5. Then C is on *three* lines: l, n, and p. **5.** Step 4

6. C is on *exactly two* lines. **6.** PF2

7. The assumption that there is less than six points leads to the contradiction that there is some point C which is on *three* lines and on *exactly two* lines; therefore, we must conclude that there are *at least* six points.

Now, from Theorem TF2 we know that the number of points in S is one of 6, 7, 8, 9, 10, . . . (at least six); that is, the number of points in S is *greater than or equal to* six. Or to say it another way, we have excluded the possibility that the number of points in S is 0, 1, 2, 3, 4, or 5. We will next establish that the number of points in S is *less than or equal to* six (at most six); that is, we will exclude the possibility that the number of points in S is 7, 8, 9, 10, ... This is proved in Theorem TF3 by assuming that the number of points in S is *at least* one greater than six and reaching a contradiction.

Theorem TF3 There are at most six points in S.

PROOF

1. There are exactly four lines, call them l, m, n, and p. **1.** PF3

2. l and m have exactly one point in common—A. **2.** PF1
   l and n have exactly one point in common—B.
   l and p have exactly one point in common—C.
   m and n have exactly one point in common—D.
   m and p have exactly one point in common—E.
   n and p have exactly one point in common—F.

3. Assume that there is more than six points;
   that is, assume that there is at least one
   point G which is distinct from A, B, C,
   D, E, and F.

   3. Assumption

4. G is on exactly two lines.

   4. PF2

5. Therefore, there are two lines which
   have more than one point in common.

   5. Steps 2 and 3

6. The assumption that there is more than
   six points (that is, the assumption that
   there is at least one point G which is distinct from A, B, C, D, E, and F) leads to
   the contradiction that there are two lines
   which have more than one point in common; therefore, we must conclude that
   there are *at most* six points.

We now use Theorems TF2 and TF3 to conclude that there are
exactly six points in the following way:

Let x be the number of points in S; then,
$x \geq 6$ (or $x \not< 6$) by Theorem TF2, and
$x \leq 6$ (or $x \not> 6$) by Theorem TF3.

Therefore, $x = 6$ by the Trichotomy (see p. 83 and reflect that
the only way x can be both less than or equal to 6 and greater than
or equal to 6 is for x to be equal to 6). Thus we have proved:

Theorem TF4 There are *exactly* six points in S.

Now we have certain information about our undefined concepts
of point and line. We know:

Points and lines exist.
There are exactly four lines.
There are exactly six points.
Every two lines have exactly one point in common.

What about the number of points on every line? Clearly none of
the four lines is an infinite set of points because there are only six
points in space! Do all of the lines contain at least one point?
Do some lines contain more points than others? These questions
are answered through a sequence of theorems.

Theorem TF5 Every line has at least one point.

PROOF

1.	There are four lines l, m, n, and p.	1.	PF3
2.	Every two lines have exactly one point in common.	2.	PF1
3.	∴ Every line contains at least one point.	3.	Step 2

Theorem TF6 Every line contains at least two points.

PROOF

1.	There are exactly four lines l, m, n, and p.	1.	PF3
2.	Assume that there is a line, say l, which has only one point, call it A.	2.	Assumption
3.	∴ A is on l and m, l and n, and l and p.	3.	PF1
4.	∴ A is on four lines and A is on exactly two lines.	4.	Step 3, PF2
5.	The assumption that there is a line with only one point leads to a contradiction; therefore, the statement		

Every line contains at least one point

is true.

Theorem TF7 Every line contains exactly three points.

PROOF

1.	There are exactly four lines l, m, n, and p.	1.	PF3
2.	There are exactly six points A, B, C, D, E, and F.	2.	TF2
3.	l contains two points, say A and C. m contains two points, say A and F. n contains two points, say E and C. p contains two points, say B and F.	3.	TF6, PF2
4.	B is on two lines. D is on two lines. E is on two lines.	4.	PF2
5.	B is on p and B is on l or on m or on n; say l.	5.	Steps 1, 3, 4

6. ∴ *l* contains at least three points *A*, *B*, and *C*.
  6. Steps 3, 5

7. Suppose *l* contains more than three points.
  7. Assumption

8. Then, *l* contains *D*, *E*, or *F*.
  8. TF4

9. Suppose *l* contains *F*.
  9. Assumption

10. Then, *F* is on at least three lines.
  10. Steps 3, 9

11. ∴ *F* is on exactly two lines and *F* is not on exactly two lines.
  11. PF2, Step 10

12. ∴ *l* does not contain *F*.
  12. *p and (not p)* is a contradiction

13. Suppose *l* contains *E*.
  13. Assumption

14. Then, *l* and *n* have two points *C* and *E* in common.
  14. Steps 3, 13

15. ∴ *l* does not contain *E*.
  (The argument to show that *l* does not contain *D* is left to the reader.)
  15. Why?

16. ∴ *E* is on *n* and *E* is on *m*, or *E* is on *n* and *E* is on *p*—Assume *E* is on *m*.
  16. Step 3, PF1, Assumption

17. Therefore, *m* contains at least three points *A*, *E*, and *F*.
  (We leave to the reader the problem of arguing that *m* contains at most three points.)
  17. Why?

18. *D* is on *n* and *D* is on *p*.
  18. Why?

19. Therefore, *n* contains at least three points *E*, *D*, and *C*, and *p* contains at least three points *B*, *D*, and *F*.
  19. Steps 3, 18

20. If *n* or *p* contains more than three points, then there is a point on more than two lines; therefore, *n* and *p* contain at most three points.

In order to get a pictorial model of our finite geometry we call special attention to the following:

PF1 Every two lines have exactly one point in common.

PF3 There are exactly four lines in *S*.

TF2 There are exactly six points in *S*.

TF7 Every line contains exactly three points.

In our model we will use an asterisk (∗) as a model of a point. A sequence of links (∞) through two or more asterisks will indicate that those points belong to the same line. A pictorial model of the finite geometry is shown in Fig. 2.1. The names of the four lines *l*, *m*, *n*, and *p*, and the names of the six points *A*, *B*, *C*, *D*, *E*, and *F* are pictured in accordance with the names adopted in the theorems.

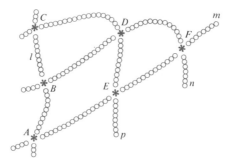

FIG. 2.1
*Model of the finite
geometry.*

We hope Fig. 2.1 helps to emphasize that the concepts of point and line in this finite geometry are different from the concepts of point and line which we have as a result of our experience with Euclidean geometry. In fact, since we defined *S* to be the set of all points, then $S = \{A,B,C,D,E,F\}$. Moreover, by Postulate PF0, each line is a subset of *S* and they are

$l = \{A,B,C\}. \qquad n = \{B,D,F\}.$
$m = \{A,E,F\}. \qquad p = \{C,D,E\}.$

The list of postulates and theorems which we have for our finite geometry now describes the system quite well. We know how points and lines are related and exactly how many of each there are. Now we had better meet head on the question of what this geometry is good for. You probably consider the Euclidean geometry which you studied in high school as being of some value because it is a fairly accurate model of the real world. Our finite geometry, since it is very small, surely cannot model the real world to the extent that Euclidean geometry does. However, it might be a model for a much different aspect of the real world. Consider the following possible, although hypothetical, situation: A group of people (points) is forming a club (set S).   It is agreed that the club will conduct its business via committees (lines).   Moreover, the club is organized in such a way that:

0. Every committee is made up of club members only (PF0).
1. Every two committees have exactly one club member in common (PF1).
2. Every person in the club is on exactly two committees (PF2).
3. There are exactly four committees (PF3).

From this information, together with the theorems of our finite geometry, we can easily conclude the following about the club:

1. Not all of the committees contain the same person.
2. The club has exactly six members.
3. Every committee has exactly three members.

If the reader will stop and reflect carefully about what we have done with our finite geometry, he may see considerable similarity between this and the way in which many professions and trades use the Euclidean geometry which you studied in school.   In a purely mathematical approach to Euclidean geometry, point and line are frequently considered undefined terms.   However, by interpreting point as the location for a bridge piling, or as the crossing of two beams, and lines as beams, an engineer can use Euclidean geometry to help him in his design of bridge structures.

Exercise set 2.1

1. For the finite geometry developed in this section make the interpretation that point corresponds to tree and line corresponds to a row, then:

   a. State Postulates PF0–PF3 using these interpretations.
   b. Draw a pictorial model of the arrangement of the trees.
   c. With these interpretations, what is the usual word for space?   (*Hint:* See Definition DF1.)
   d. What statements can you make about the arrangement of this orchard?

2. Use telephone and telephone line to make a model of the finite geometry developed in this section.

   a. Using the interpretation that a point is a telephone and a line is a telephone line, restate Postulates PF0–PF3.

    b. Draw a pictorial model of the arrangement of telephones.

    c. With this interpretation, what is the usual word for space? (*Hint:* See Definition DF1.)

    d. Prove: Not all telephone lines contain the same telephone.

3. Consider the finite geometry with the following:

Undefined term: point.
Postulate 1. There are at least two points.
Definition 1. The set S of all points is called space.
Definition 2. Every two-point subset of S is called a line; that is, every two points are on exactly one line.
Postulate 2. For every line, there is a point not on the line.
Definition 3. Two lines are parallel if and only if they have no point in common.
Postulate 3. Given a line and a point not on the line, there exists exactly one line parallel to the given line which contains the given point.

    a. Draw a pictorial model of this geometry similar to that in Fig. 2.1.

    b. Prove: There are exactly four points in S.

    c. Prove: Each point is contained on at least two lines. (Give two proofs: One a direct proof using part b and another a proof by contradiction.)

4. Consider a mathematical system composed of the following:

Undefined terms: bead, wire, bead on a wire.
Definition: The set S of all beads and wires is called an *ornament*.
Postulates:

    (1) Every two wires have at least one bead in common.
    (2) Every two wires have at most one bead in common.
    (3) Every bead in S is on at least two wires.
    (4) Every bead in S is on at most two wires.
    (5) There are exactly four wires in S.

    a. Prove: Every two wires in S have exactly one bead in common.

    b. Prove: Every bead in S is on exactly two wires.

    c. How many beads are in S?

    d. Draw a model of the ornament.

e.   How many beads are there on each wire?

f.   Parallel beads are defined as follows:

Beads are *parallel* if and only if they are not on a common wire.

(1)   List the parallel beads.

(2)   Is the following statement true or false?

Every bead in S has exactly one bead parallel to it.

5.   Consider a mathematical system composed of the following:

Undefined terms: politician, committee, committee contains a politician (or a politician is on a committee).
Definition: S is the set of all politicians.
Postulates:

(1)   Every two politicians in S are on at least one committee.

(2)   Every two politicians in S are on at most one committee.

(3)   Every two committees have at least one politician in common.

(4)   There exists at least one committee.

(5)   Every committee contains at least one politician.

(6)   Not all politicians are on the same committee.

(7)   No committee contains more than three politicians.

a.   How many politicians are in S?

b.   Draw a model which will suggest the committee arrangements.

c.   Does there exist a committee which contains more than three politicians?

d.   Does there exist a committee which contains less than three politicians?   Explain your answer.

6.   Consider the mathematical structure composed of the following:

Undefined terms: nail, board, nail in a board.
Definition: The set of all nails and boards is called a *design*.
Postulates:

(1)   Every two boards have at least one nail in common.

(2)   Every two boards have at most one nail in common.

(3)   Every nail in S is in at least two boards.

(4) Every nail in S is in at most two boards.
(5) The total number of boards in S is four.

  a. How many boards are there in S?
  b. How many nails are there in S?
  c. How many nails are there in each board?
  d. Draw a model of the design.
  e. Prove: Every nail in S is in exactly two boards.

2.3 A DEDUCTIVE INTRODUCTION
TO EUCLIDEAN GEOMETRY

The etymology of the word "geometry" indicates that this branch of mathematics has its historical origin related to measurement of distances on the earth. The geometry which we usually study in elementary and secondary schools is called "Euclidean geometry." The term "Euclidean geometry" is used to honor the Greek mathematician and philosopher Euclid, who about 300 B.C. made the first attempt to place the study of geometry on a firm logical or deductive basis.

Since geometry, at the time of Euclid, was in fact related to spatial relationships and measurement of distances on the earth, it is not surprising that our everyday experiences serve us rather well in understanding some of the basic ideas of Euclidean geometry. However, to illustrate how geometry can be developed from certain basic assumptions, we will begin our study of Euclidean geometry in this formal manner, and progress in this spirit far enough to illustrate a formal development and to suggest how extensive an undertaking it is to develop all the concepts in this way. We will prove, from our basic assumptions, some notions about points, lines, and planes which we usually take for granted.

We began our study by listing our *undefined terms*:

1. Point
2. Line
3. Plane.

Definition 2.1 *Euclidean space* (or simply *space*), denoted by S, is the set of all points.

Some basic assumptions which we make about the undefined concepts are the following postulates:

Postulate 2.1   S is an infinite set of points.

Postulate 2.2   A line is an infinite set of points.

Postulate 2.3   For every two points in space there exists at least one line which contains them.

Postulate 2.4   For every two lines, their intersection set contains exactly one point or is the empty set.

Postulate 2.5   For every line $\mathscr{L}$ there exists a point $A \in S$ such that A is not on $\mathscr{L}$.

Theorem 2.1   For all points $A, B \in S, A \neq B$, there exists exactly one line $\mathscr{L}$ such that $\{A,B\} \subseteq \mathscr{L}$. That is, through every two points in S there is exactly one line.

PROOF

1. $A, B \in S, A \neq B$.	1. Given
2. Therefore, there is *at least* one line, call it $\mathscr{L}$, such that $\{A,B\} \subseteq \mathscr{L}$.	2. Postulate 2.3
3. Now suppose $\mathscr{M}$ is a line such that $\{A,B\} \subseteq \mathscr{M}$ and $\mathscr{M} \neq \mathscr{L}$.	3. Assumption
4. $\{A,B\} \subseteq \mathscr{M} \cap \mathscr{L}$.	4. p. 23
5. $\mathscr{M} \cap \mathscr{L} = \varnothing$ or $\mathscr{M} \cap \mathscr{L}$ contains exactly one point.	5. Postulate 2.4
6. $\therefore A = B$.	6. Postulate 2.4, Step 4
7. The assumption that there is more than one line which contains A and B leads to the contradiction that $A \neq B$ and $A = B$; therefore, we must conclude that there is *at most* one line which contains A and B.	7. Steps 1, 5
8. $\therefore$ There is exactly one line which contains A and B.	

Theorem 2.1 is often stated as: Every two points determine a line.

Let us think a little bit about the significance of Theorem 2.1. That is, let us look at what it tells us about a line. To draw models we will use a dot (·) as a model of a point and a path (~) as a model of a line. What Theorem 2.1 tells us is that the sort of thing suggested in Fig. 2.2 cannot happen where $\mathscr{L}$ and $\mathscr{M}$ are both lines.

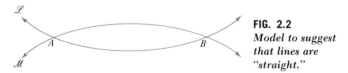

FIG. 2.2
Model to suggest
that lines are
"straight."

Thinking about Theorem 2.1 further with reference to the model in Fig. 2.2, we can see that the theorem really describes what a model of a line must look like; that is, it must have the physical property which we refer to as straightness. Therefore, in order for a path to be a model of a line, it should look like the model in Fig. 2.3.

FIG. 2.3
Model of a line.

Now we are beginning to develop some ideas about lines: We postulated (that is, assumed) that a line is an infinite set of points. Theorem 2.1 gives us an additional characterization of a line—it has the property we call "straightness." What about the question of whether the set of all lines is an infinite set? We answer this question in Theorem 2.2.

Theorem 2.2 S contains an infinite set of lines.

PROOF

1. There exists a line, call it $\mathscr{L}$.	1. Theorem 2.1
2. There exists a point in space not on $\mathscr{L}$, call it A.	2. Postulate 2.5
3. $\mathscr{L}$ is an infinite set of points.	3. Postulate 2.2
4. For every point X on $\mathscr{L}$, X and A determine a line.	4. Theorem 2.1
5. Assume that there exist two points X_1 and X_2 on $\mathscr{L}$ such that $\overleftrightarrow{AX_1} = \overleftrightarrow{AX_2}$.	5. Assumption
6. $\therefore X_1 \in \overleftrightarrow{AX_2}$.	6. Concept of $=$ for sets

7.	$\therefore \{X_1, X_2\} \subseteq \mathcal{L}$.	7.	Definition of $\subseteq$
8.	$\therefore \overleftrightarrow{AX_2} \subseteq \mathcal{L}$.	8.	Theorem 2.1
9.	$\therefore A \in \mathcal{L}$.	9.	Definition of $\subseteq$
10.	The assumption that two points on $\mathcal{L}$ determine with point A a common line leads to the contradiction "$A \notin \mathcal{L}$ and $A \in \mathcal{L}$"; therefore, S contains an infinite set of lines.		

Now let us turn our attention to the undefined concept of plane. What assumptions should we make about a plane and what properties can we deduce about planes? The assumptions that are made vary from one author to another. We make assumptions about planes which are analogous to those we made about lines. We need, first of all, the concept of collinear points.

Definition 2.2   A set of points $\mathcal{A}$ is *collinear* if and only if there exists one line which contains every point of $\mathcal{A}$.

If any element of a set of points does not belong on the same line as other points in the set, we call it a *noncollinear set*. Our concern with collinear and noncollinear sets will most frequently be with sets of three points.

Postulate 2.6   For every three noncollinear points there exists at least one plane which contains them.

Postulate 2.7   For every two planes, the intersection set is a line or is the empty set.

Definition 2.3   For all lines $\mathcal{L}$ and $\mathcal{M}$ in a plane α, $\mathcal{L}$ is *parallel* to $\mathcal{M}$ (we write $\mathcal{L} \parallel \mathcal{M}$) if and only if $\mathcal{L} = \mathcal{M}$ or $\mathcal{L} \cap \mathcal{M} = \emptyset$.

Theorem 2.3   The relation $\parallel$ on the set of all lines in a plane has the following properties.

1. For every line $\mathcal{L}$, $\mathcal{L} \parallel \mathcal{L}$.
2. For all lines $\mathcal{L}$ and $\mathcal{M}$, if $\mathcal{L} \parallel \mathcal{M}$, then $\mathcal{M} \parallel \mathcal{L}$.
3. For all lines $\mathcal{L}$, $\mathcal{M}$, and $\mathcal{N}$, if $\mathcal{L} \parallel \mathcal{M}$ and $\mathcal{M} \parallel \mathcal{N}$, then $\mathcal{L} \parallel \mathcal{N}$.

The proof is left as an exercise for the reader.

Definition 2.4   For all planes α and β, α is *parallel* to β (we write $\alpha \parallel \beta$) if and only if $\alpha = \beta$ or $\alpha \cap \beta = \varnothing$.

Definition 2.5   For every line $\mathscr{L}$ and for every plane α, $\mathscr{L}$ is *parallel* to α (we write $\mathscr{L} \parallel \alpha$) if and only if $\mathscr{L} \subseteq \alpha$ or $\mathscr{L} \cap \alpha = \varnothing$.

Theorem 2.4   The relation $\parallel$ on the set of all planes has the following properties:

1.   For every plane α, $\alpha \parallel \alpha$.
2.   For all planes α and β, if $\alpha \parallel \beta$, then $\beta \parallel \alpha$.
3.   For all planes α, β, and γ, if $\alpha \parallel \beta$ and $\beta \parallel \gamma$, then $\alpha \parallel \gamma$.

The proof is left as an exercise for the reader.

Postulate 2.8   For every line and for every plane, there exists a point in space not on the line or on the plane.

Postulate 2.9   Every plane contains at least one line and at least one point not on the line.

We leave to the reader the problem of proving that there do in fact exist three points which are noncollinear, and hence that there exists at least one plane (Exercise 1 of Exercise Set 2.2).

Theorem 2.5   For all A, B, $C \in S$ which are noncollinear points, there exists exactly one plane which contains them.

The proof parallels to a great extent the proof of Theorem 2.1 and is left as an exercise for the reader.

The statement of Theorem 2.5 is often given as:

Every three noncollinear points determine a plane.

Theorem 2.5 is to planes as Theorem 2.1 is to lines.   Recall that Theorem 2.1 intuitively states that a line has a property which we call straightness.   Similarly, Theorem 2.5 tells us that a plane has the property which we refer to as flatness.   That is, it tells us that the situation suggested in Fig. 2.4 cannot happen for two planes α and β.

In order for a physical object or a picture to be a model of a plane, it must exhibit the physical property of being flat.   Thus,

FIG. 2.4
*Model to suggest
that planes are
"flat."*

we think of a table top or a wall as being a model of a plane because
they exhibit this property.   We will use a picture as shown in Fig.
2.5 to suggest, or to be a model of a plane.

FIG. 2.5
Model of a plane.

There are many more questions that need to be answered about
lines, planes, and space.   Most of these we delay to the next chapter where we will discuss them in a less formal manner than we
are doing in this section.

Intuitively, the ideas of straightness and flatness are very similar.
The idea of straightness is related to "one dimension" as flatness
is related to "two dimensions." The two concepts are certainly
closely related in a physical real-world sense. We prove in Theorem 2.6 that line and plane are related in Euclidean geometry in
an analogous manner.

Theorem 2.6 For every plane α, if α contains two points A and
B of a line $\mathscr{L}$, then $\mathscr{L} \subseteq \alpha$.

PROOF

1. α is a plane, $\mathscr{L}$ is a line, $\{A,B\} \subseteq \alpha$, and   1. Why?
$\{A,B\} \subseteq \mathscr{L}$.

2.	$\therefore \{A,B\} \subseteq \alpha \cap \mathcal{L}$.	**2.** ?
3.	There exists a point C in space such that $C \notin \alpha$ and $C \notin \mathcal{L}$.	**3.** ?
4.	There exists a plane β which contains the the points A, B, and C.	**4.** ?
5.	$\alpha \cap \beta$ is a line or $\alpha \cap \beta = \varnothing$.	**5.** ?
6.	$\{A,B\} \subseteq \alpha \cap \beta$.	**6.** ?
7.	$\alpha \cap \beta \neq \varnothing$.	**7.** ?
8.	$\therefore \alpha \cap \beta$ is a line.	**8.** ?
9.	$\therefore \alpha \cap \beta = \mathcal{L}$.	**9.** ?
10.	$\therefore \mathcal{L} \subseteq \alpha$.	**10.** ?

Theorem 2.6 further confirms our notion that lines can be characterized as being straight and planes can be characterized as being flat. Or at least, lines and planes can be characterized as being straight and flat, respectively; or else, neither lines nor planes can be characterized as being straight or flat, respectively. This is because Theorem 2.6 disallows both of the situations suggested in Fig. 2.6.

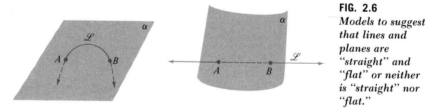

FIG. 2.6
Models to suggest that lines and planes are "straight" and "flat" or neither is "straight" nor "flat."

All of our postulates and theorems indicate that the picture shown in Fig. 2.7 is an appropriate model to indicate the situation of Theorem 2.6.

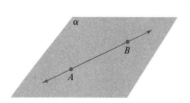

FIG. 2.7
Model to suggest that if a plane contains two points, then it contains the line determined by the points.

We are beginning to "pin down" the relationship between points, lines, and planes. However, there are still some questions worth looking at and making sure that the questions raised follow from our assumptions. It is an easy argument to show that

Every point in space is on at least one line

is deducible from facts which we have assumed or proved; we leave to the reader the problem of thinking this through.

Two related questions to which we should like to give attention are the following:

Is every point contained in some plane?
Is every line contained in some plane?

Be careful, the answers to these questions seem obvious; but, what we want to know is whether or not we can establish these as facts from our assumptions. The answers to both questions are presented in Theorem 2.7.

Theorem 2.7 For every point A in space and for every line $\mathscr{L}$, there exists a plane which contains A and $\mathscr{L}$.

We leave the proof of this theorem as an exercise for the reader.

2.4 GENERAL COMMENTS AND EXERCISES

Before we leave this chapter for a more informal study of geometry, let us look back over Sections 2.2 and 2.3. The finite geometry in Section 2.2 (in addition to demonstrating the structure of mathematics) as compared with the Euclidean geometry begun in Section 2.3 shows what different concepts arise when different basic assumptions are made about the undefined terms. It really underscores how important the postulates are in developing additional characteristics and relationships among the undefined terms. No doubt you had some difficulty in the finite geometry letting yourself think of a line as a set containing exactly three points— our past experience screams at us that this is not true! On the other hand, we need to keep in mind that our finite geometry (although maybe a little artificially) turned out to be a *mathematical model* of real-world situations with appropriate physical interpretations of the undefined concepts. In Section 2.3, we began with some basic assumptions about our undefined concepts of point, line, and plane and were able to deduce formally some descriptive characteristics of line and plane which agree with physical attributes of objects which we frequently use to suggest the concept of line and plane. In a limited sense this demonstrates some of the activities of a mathematician; he is willing to allow the real world to suggest certain things in mathematics, but he is not willing to accept these as part of his mathematical system until he has proved them from his basic assumptions.

Exercise set 2.2

1. Prove: There are at least three points which are noncollinear.

2. a. Draw pictures which represent the three alternatives in the following statement:

   The intersection set of a line and a plane is a set which contains exactly one point, is the empty set, or is a line.

   b. Outline the proof of the statement in part a.

3. Prove Theorem 2.5.

4. Prove: If A, B, and C are any three points, then there is a plane which contains them.

5. Prove: Two lines whose intersection set is nonempty determine a plane.

6. Prove: A plane α is an infinite set of points.

7. Prove: For all planes α and β, if $\alpha \cap \beta \neq \emptyset$, then $\alpha \cap \beta$ is an infinite set.

8. Prove: Every plane α contains an infinite set of lines.

9. Prove: S contains an infinite set of planes. (*Hint:* Use Exercise 8 and Postulate 2.8.)

3

subsets of lines

3.1 INTRODUCTION

In this chapter we will use the concept of a relation defined on space called the *betweenness* relation. From a mathematical point of view we should, therefore, give a definition of this relation before we proceed; however, its definition depends on the concepts of distance and measure, and for pedagogical reasons, we wish to delay this until later. For the present we will proceed with a naive notion of the betweenness relation which is suggested by the models in Fig. 3.1.

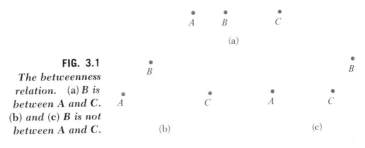

FIG. 3.1
The betweenness relation. (a) *B is between A and C.* (b) *and* (c) *B is not between A and C.*

The first requirement for a point B to be between A and C is that A, B, and C be collinear; the other requirement (and a formal definition) we will give when we have considered distance and measure. To indicate that B is between A and C we write A–B–C.

3.2 LINES AND LINE SEGMENTS

We proved in Chapter 2 that for every line there exists a plane which contains it; therefore, a line is a plane figure. We also proved in Chapter 2 that for every A and B which are distinct points in S there is *exactly* one line which contains these two points. We say that the points A and B *determine* the line which contains them. If A and B are two points, the line which contains A and B (or is determined by A and B) is denoted by $\overleftrightarrow{AB}$ or $\overleftrightarrow{BA}$. We emphasize: For all $A, B \in S$, $\overleftrightarrow{AB}$ is a *set*, a set of points. Moreover, for all $A, B \in S$, $\overleftrightarrow{AB}$ is an *infinite set* of points. We will consistently use a model as shown in Fig. 3.2 as the model for the line

Fig. 3.2
Line determined by A and B.

which contains (or is determined by) A and B. The arrowheads are to suggest that the model is to continue in both directions *without end*. Note the significance of the phrase "without end"; it tells us that a line has *no endpoints*.

We adopted early in our study the basic notion of equality; however, for emphasis, we review the meaning of equality in the setting of geometry. We shall continue to use $A, B, C, \ldots$ as names for points. Thus, when we write $A = B$, we mean "A" and "B" are names for the same point and when we write $\overleftrightarrow{AB} = \overleftrightarrow{BA}$, we mean that "$\overleftrightarrow{AB}$" and "$\overleftrightarrow{BA}$" are names for the same *set of points* (line, in this case). More generally, to write $\overleftrightarrow{AB} = \overleftrightarrow{CD}$ means that "$\overleftrightarrow{AB}$" and "$\overleftrightarrow{CD}$" are names for the same line. If $\overleftrightarrow{AB} = \overleftrightarrow{CD}$, then it follows that $A \in \overleftrightarrow{CD}$ and $B \in \overleftrightarrow{CD}$; that is, A, B, C, and D are collinear. There are many possible betweenness relationships for A, B, C, and D when $\overleftrightarrow{AB} = \overleftrightarrow{CD}$, one of which is suggested in the model in Fig. 3.3.

The concept of *line segment* is very important in geometry. We present the definition of line segment and then make observations about line segments in view of the definition.

FIG. 3.3
*A betweenness
relation for A, B,
C, D.*

Definition 3.1 For all points $A, B \in \mathscr{L}$, $A \neq B$, the subset of $\mathscr{L}$ which contains A, B, and all points between A and B, is the *line segment* determined by A and B; A and B are called *endpoints*. We denote this set by $\overline{AB}$ (read "line segment AB"). That is, $\overline{AB} = \{X \in \mathscr{L} \mid X = A, X = B, \text{ or } A\text{–}X\text{–}B\}$.

The following are very easy observations from Definition 3.1.

1. A line segment is a set of points.
2. A line segment is a subset of a line; and therefore, can also be characterized as being straight.
3. $A \in \overline{AB}$ and $B \in \overline{AB}$; that is, every line segment contains its endpoints.
4. $\overline{AB}$ is an infinite set of points; from this it follows that a set which contains only one point is not a line segment. Thus, "$\overline{AA}$" does not name a line segment.
5. For every point $X \in S$, $X \in \overline{AB}$ if and only if $X = A$, $X = B$, or $A\text{–}X\text{–}B$.

We will refer to $\overline{AB}$ in various ways: namely, "the line segment AB," "the line segment determined by A and B," or "the line segment from A to B." Every line is an infinite set of points and, therefore, the set which contains all possible *pairs* of points on a line is infinite. Since every pair of points determines a line segment, it follows that the set consisting of all line segments of a line is an infinite set.

Definition 3.2 For all line segments AB and CD, $\overline{AB}$ is *parallel* to $\overline{CD}$ if and only if $\overleftrightarrow{AB} \parallel \overleftrightarrow{CD}$. If $\overline{AB}$ is parallel to $\overline{CD}$, we write $\overline{AB} \parallel \overline{CD}$.

The preceding concepts and terminology are illustrated in Examples 3.1 and 3.2.

EXAMPLE 3.1 Let points $A, B, \ldots, H$ and line segments determined by them be related as suggested in the accompanying

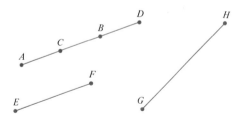

model.   The following are true statements:

a. $\overline{EF} \parallel \overline{AB}.$ b. $\overline{EF} \nparallel \overline{GH}.$ c. $\overline{AB} \parallel \overline{CD}.$
d. $\overline{EF} \parallel \overline{EF}.$ e. $\overline{AC} \parallel \overline{CB}.$ f. $\overline{AC} \parallel \overline{BD}.$

EXAMPLE 3.2 Consider the line which contains points A, B, C, and D, which are related as suggested in the accompanying model.   Each of the following is a true statement.

$$A \qquad\qquad C \quad D \quad B$$

a. $\overleftrightarrow{AC} = \overleftrightarrow{DB}.$ b. $\overleftrightarrow{CD} = \overleftrightarrow{AB}.$
c. $\overleftrightarrow{BA} \subseteq \overleftrightarrow{CD}.$ (Think!)   d. $\overline{CD} \subseteq \overleftrightarrow{AB}.$
e. $\overline{CD} \subseteq \overline{AB}.$ f. $\overline{AD} \cap \overline{CB} = \overline{CD}.$
g. $\overline{CD} \cap \overline{DB} = \{D\}.$ h. $\overline{AC} \cup \overline{CD} = \overline{AD}.$
i. $D \in \overline{CB}.$ j. $A \notin \overline{CD}.$
k. $A \in \overleftrightarrow{CD}.$ l. $A \in \overleftrightarrow{DB}.$
m. $\overleftrightarrow{AC} \parallel \overleftrightarrow{AD},$ n. $\overline{AC} \parallel \overline{AD}.$
o. $\overline{AC} \parallel \overline{DB}.$

Exercise set 3.1

1. Let lines $\mathscr{L}_1$ and $\mathscr{L}_2$ contain specific points as suggested in the accompanying model.   Indicate whether the following are true or false.

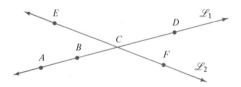

a. A–E–D.
(Read: E is between A
and D.)

b. E–C–F.

c. F–C–E.

d. $\overleftrightarrow{BC} \cap \overleftrightarrow{EF} = \{C\}$.

e. $\overline{BD} \cap \overline{CF} = \varnothing$.

f. $\overline{BC} \subseteq \overline{BD}$.

g. $\overline{BC} \cup \overline{CD} = \mathcal{L}_1$.

h. $\overleftrightarrow{AB} = \mathcal{L}_1$.

i. $\overleftrightarrow{AB} \cap \overline{AB} = \{A,B\}$.

j. $\{A,B\} \subseteq \overline{AB}$.

k. $\overleftrightarrow{AB} \parallel \overleftrightarrow{EF}$.

l. $\overleftrightarrow{AB} \parallel \overleftrightarrow{CD}$.

m. $\overline{AB} \parallel \overline{CD}$.

n. $\overline{AB} \cap \overline{BC} = B$.

o. $\overline{BC} \subseteq \overline{CB}$.

p. $CB \subseteq \overline{BC}$.

q. $\overline{CB} = \overline{BC}$.

r. $C \in \mathcal{L}_1 \cap \mathcal{L}_2$.

s. $B \in \overline{CD} \cup \overline{CF}$.

t. $B \in \overline{AC} \cup \overline{CF}$.

u. $\overline{BC} = \overline{CD}$.

v. $\overline{AB} = \overline{BC}$.

w. $\overline{AB}$ is a collinear set.

x. $\{B,C,F\}$ is a collinear set.

y. $\overline{BC} \cup \overline{CF} \subset \mathcal{L}_1 \cup \mathcal{L}_2$.

z. $\overline{EC} \cup \overline{CB} = \overline{DC} \cup \overline{CF}$.

A. $\overline{BC} \cup \overline{CD} = \overline{BD}$.

B. $\overleftrightarrow{AB} \cap \overleftrightarrow{CD} = \overleftrightarrow{AD}$.

C. $\overleftrightarrow{AB} \cap \overleftrightarrow{EF} = \varnothing$.

2. Let points, line segments, and lines which are labeled in the accompanying model be related *as suggested by the model*. Complete each of the following to make true statements.

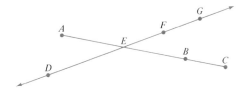

a. $\overline{DE} \cap \overline{FG} = ?$.

b. $\overline{AC} \cap \overline{DG} = ?$.

c. $\overline{AC} \cap \overline{BC} = ?$.

d. D–?–F.

e. $\overline{DF} \cup \overline{EG} = ?$.

f. $\overline{DF} \cap \overline{EG} = ?$.

g. $\overline{AC} \cap \overline{EF} = ?$.

h. $\overleftrightarrow{EF} \cap \overleftrightarrow{FG} = ?$.

i. $\overline{EF} \cap \overline{FG} = ?$.

j. $\overline{AB} \subseteq ?$.

k. $\overline{DE} \cup \overline{EG} = ?$.

l. $\overleftrightarrow{DE} \cup \overleftrightarrow{EG} = ?$.

3. Let $\overline{AD}$ and $\overline{BC}$ be line segments as suggested by the accompanying model. Show a diagrammatic scheme which suggests that there is a one-to-one correspondence between $\overline{AD}$ and $\overline{BC}$.

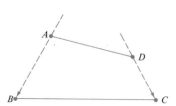

4. Draw a model which will suggest that there is a one-to-one correspondence between a line segment, except its endpoints, and a line.

5. Draw a model of a plane figure which suggests each of the following:

    a. $\overleftrightarrow{AB} \cap \overleftrightarrow{CD} = \varnothing$.   **b.** $\overleftrightarrow{AB} \cap \overleftrightarrow{CD} = \{P\}$.

    c. $\overleftrightarrow{AB} \cap \overleftrightarrow{CD} = \{C\}$.   **d.** $\overline{AB} \cap \overleftrightarrow{AB} = \overline{AB}$.

    e. $\overline{AB} \cap \overleftrightarrow{CD} = \{A\}$.   **f.** $\overline{AB} \cap \overline{CD} = \overline{CD}$.

    g. $\overleftrightarrow{AB} \cap \overleftrightarrow{CD} = \overleftrightarrow{CD}$.   **h.** $\overline{AB} \cup \overline{BC} = \overline{AC}$.

    i. $\overline{AB} \cup \overline{BC} \neq \overline{AC}$.   **j.** $\overline{AB} \cup \overline{CD} = \overline{AD}$.

    k. $\overline{AB} \subseteq \overleftrightarrow{CD}$.   **l.** $\overline{AB} \subseteq \overline{EF}$.

    m. $\overline{AB} \nsubseteq \overline{AC}$.   **n.** $\overline{AB} \cap \overline{BC} = \{B\}$.

    o. $\overline{AB} \cap \overline{CD} = \varnothing$ and $\overleftrightarrow{AB} \cap \overleftrightarrow{CD} \neq \varnothing$.

6. Prove: If A and B are two points in plane α, then α contains every point of $\overline{AB}$ (that is, $\overline{AB} \subseteq \alpha$).

7. We frequently refer to physical objects as being to the right or to the left of a given object. Moreover, in your past experience with geometry you have undoubtedly referred to a point as being to the left or to the right of a given point. The model below suggests what one has in mind in saying B is to the right of A. Although we often use the words "right" and "left" in

geometry, they are not precise geometric notions.   Using the model, set builder notation, and the concept of betweenness, describe the set of *all* points which one would consider as being to the right of *A*.

3.3 CONVEX SETS, SEPARATION OF LINES, MORE SUBSETS OF LINES

Another important concept which is useful in geometry is that of *convex set.*   As we continue our study of Euclidean geometry we will find many examples of sets of points (geometric figures) which are convex sets.   However, at this point we have only a few such sets to illustrate the concept.   First, consider a line $\mathscr{L}$ (a set): For every two points *A*, *B* on $\mathscr{L}$, how is the line segment *AB* related to $\mathscr{L}$?   Clearly $\overline{AB} \subseteq \mathscr{L}$; that is, *for every two points on a line $\mathscr{L}$, the line segment determined by the two points is a subset of $\mathscr{L}$.* Similarly for a line segment *AB*, every two points on $\overline{AB}$ *determines a line segment which is a subset of* $\overline{AB}$.   The same is true for planes and space; that is, for every *A* and *B* which are points in a plane α, $\overline{AB} \subseteq \alpha$, and for every *A* and *B* which are points in space, $\overline{AB} \subseteq S$. Lines, line segments, planes, and space are *convex sets;* the characterizing idea of *convex set* is that, *for every two points in the set, the line segment determined by the two points is a subset of the set.*

Definition 3.3   For every set $\mathscr{A}$, $\mathscr{A} \subseteq S$, $\mathscr{A}$ is a *convex set* if and only if for every $X, Y \in \mathscr{A}$, $X \neq Y$, $\overline{XY} \subseteq \mathscr{A}$.

An illustration is given in Example 3.3.

EXAMPLE 3.3
a.   The accompanying model is of $\overline{AB} \cup \overline{BC}$, where *A*, *B*, and *C* are collinear points and *X* and *Y* are elements of $\overline{AB} \cup \overline{BC}$.

(1)   Do "*X*" and "*Y*" name points?
(2)   Do *X* and *Y* determine a line segment?
(3)   Is the line segment *XY* a subset of $\overline{AB} \cup \overline{BC}$; that is, is $\overline{XY} \subseteq \overline{AB} \cup \overline{BC}$?
(4)   Is $\overline{AB} \cup \overline{AC}$ a convex set?

b. The accompanying model is of $\overline{AB} \cup \overline{BC}$, where A, B, and C are noncollinear points and X and Y are elements of $\overline{AB} \cup \overline{BC}$.

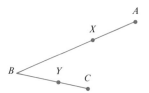

(1) Do "X" and "Y" name points?
(2) Do X and Y determine a line segment?
(3) Is the line segment XY a subset of $\overline{AB} \cup \overline{BC}$; that is, is $\overline{XY} \subseteq \overline{AB} \cup \overline{BC}$?
(4) Is $\overline{AB} \cup \overline{BC}$ a convex set?

Another idea which we will find useful in defining special sets of points is that of separation. For every point B on a line $\mathscr{L}$, B *separates* the line into two sets; namely, the set of points which we intuitively think of as being to the right of B, and the set of points which we intuitively think of as being to the left of B. It seems apparent that the sets of points to the right and to the left, respectively, are both convex sets. We illustrate with a model in Example 3.4 and then postulate the precise idea.

EXAMPLE 3.4 Let A, B, and C be points on line $\mathscr{L}$ and be related as suggested in the accompanying model. The point B *separates* line $\mathscr{L}$ into two sets:

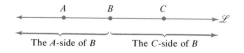

the set of points "to the right of B" which is called a *half-line*,

and

the set of points "to the left of B" which is called a *half-line*.

The half-lines "to the right of B" and "to the left of B" are called *opposite* half-lines.

Consider the following questions.

a. Does it appear that each half-line is a convex set?
b. Is B an element of either half-line?
c. Is the union of two opposite half-lines a line?

Postulate 3.1 (*Line separation postulate*) For every line $\mathscr{L}$ and for every point X on $\mathscr{L}$, X separates the points on $\mathscr{L}$ not equal to X into two convex sets such that if P is in one set and Q is in the other, then $\overline{PQ} \cap \{X\} \neq \varnothing$.

Definition 3.4   For every line $\mathscr{L}$ and for every point X on $\mathscr{L}$, the two convex sets determined by X are called *half-lines;* the two half-lines determined by X are called *opposite* half-lines, and X is called the *boundary point* of each half-line.

If $P \in \overline{AB}$, the set of points on $\overleftrightarrow{AB}$ which we intuitively describe as being "to the right of P" is called the "half-line PB" and is denoted by $\overset{\circ\to}{PB}$.   We leave to the reader the problem of intuitively describing $\overset{\circ\to}{PA}$.

In the symbol $\overset{\circ\to}{PB}$, the circle (o) indicates that the point P is not a point of the set; the arrow ($\to$) indicates that the set includes all of the points which are intuitively described as being "on the same side of P as B."

The model in Fig. 3.4 illustrates the concept of a half-line AB and a formal definition of half-line is given in Definition 3.5.

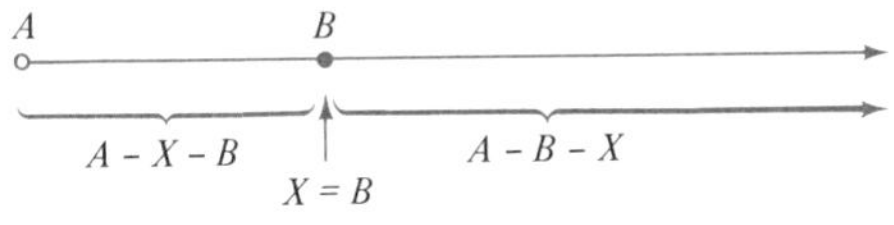

FIG. 3.4
Half-line AB.

Definition 3.5   For every A, $B \in \mathscr{L}$, the *half-line* determined by A and B, denoted by $\overset{\circ\to}{AB}$, is defined as

$$\overset{\circ\to}{AB} = \{X \in \mathscr{L} | A\text{–}X\text{–}B, X = B, \text{ or } A\text{–}B\text{–}X\}.$$

An idea which is useful in developing further geometric concepts is that of a ray.   The idea of ray is similar to that of half-line; in fact, the set $\overset{\circ}{AB} \cup \{A\}$ is called the "ray AB with endpoint A"; we denote it by $\overrightarrow{AB}$. The model in Fig. 3.5 illustrates the concept of a ray AB.

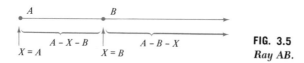

$A - X - B$ $A - B - X$ **FIG. 3.5**
$X = A$ $X = B$ *Ray AB.*

Definition 3.6 For every two points A and B, the *ray* with endpoint A and containing B is defined as

$$\overrightarrow{AB} = \{A\} \cup \overset{\circ}{AB}.$$

The following statements are equivalent definitions for ray AB.

$$\overrightarrow{AB} = \overline{AB} \cup \{X \in \overleftrightarrow{AB}|A\text{–}B\text{–}X\},$$
$$\overrightarrow{AB} = \{X \in \overleftrightarrow{AB}|X = A,\ A\text{–}X\text{–}B,\ X = B,\ \text{or}\ A\text{–}B\text{–}X\}.$$

If D separates line AB into two half-lines, $\overset{\circ}{DA}$ and $\overset{\circ}{DB}$, then $\overrightarrow{DA}$ and $\overrightarrow{DB}$ are called *opposite* rays.

Some illustrations are given in Example 3.5.

EXAMPLE 3.5 Consider the line $\mathscr{L}$ with specific points as suggested in the accompanying model. The following are true statements and illustrate some of the relationships among points, half-lines, rays, line segments, and lines.

A B C D $\mathscr{L}$

a. $\overset{\circ}{BA} \cap \overset{\circ}{BC} = \varnothing.$ **b.** $\overset{\circ}{BA} \cap \overrightarrow{BC} = \varnothing.$

c. $\overset{\circ}{BA} \cup \overrightarrow{BC} = \overleftrightarrow{AB}.$ **d.** $\overrightarrow{BA} \cup \overrightarrow{BC} = \overleftrightarrow{AB}.$

e. $\overset{\circ}{BA} \cup \overset{\circ}{BC} \neq \overleftrightarrow{AB}.$ **f.** $A \in \overrightarrow{AB}.$

g. $A \notin \overset{\circ}{AB}.$ **h.** $\overrightarrow{BC} \cap \overrightarrow{CB} = \overline{BC}.$

i. $\overset{\circ}{BC} \cap \overset{\circ}{CB} = \overline{BC} \setminus \{B,C\}.$ **j.** $\overset{\circ}{BC} \cap \overrightarrow{CB} = \overline{BC} \setminus \{B\}.$

Exercise set 3.2

1. Let $\mathscr{L}$ be a line and B a point on $\mathscr{L}$. B separates $\mathscr{L}$ into two half-lines; let A and C be on different half-lines of $\mathscr{L}$. Indicate whether the following are true or false. (*Hint:* Draw a model.)

a. $\overleftrightarrow{AC} \subseteq \mathscr{L}$.
b. $\overset{\circ}{\overrightarrow{BC}} \subseteq \mathscr{L}$.
c. $\overset{\circ}{\overrightarrow{BC}} \cap \overset{\circ}{\overrightarrow{BA}} = \{B\}$.
d. $B \in \overset{\circ}{\overrightarrow{BC}}$.
e. $B \in \overset{\circ}{\overrightarrow{BA}}$.
f. $B \in \overline{AC}$.
g. $\overrightarrow{BC} \subseteq \mathscr{L}$.
h. $\overrightarrow{CB} \subseteq \mathscr{L}$.
i. $C \in \overrightarrow{AB}$.
j. $C \in \overrightarrow{BA}$.
k. $B \in \overrightarrow{BC}$.
l. $B \in \overrightarrow{BA}$.
m. $\{B,C\} \subseteq \overrightarrow{AC}$.
n. $\{A,B,C\} \subseteq \overrightarrow{AC}$.
o. $\{B\} \subseteq \overrightarrow{AC}$.

2. From the accompanying model indicate whether each statement below is true or false.

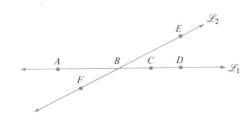

a. $\overset{\circ}{\overrightarrow{BC}} \cup \overset{\circ}{\overrightarrow{CD}} = \overrightarrow{DC}$.
b. $\overset{\circ}{\overrightarrow{BC}} \cap \overset{\circ}{\overrightarrow{BE}} = \{B\}$.
c. $\overrightarrow{BC} \cap \overrightarrow{BE} = B$.
d. $\overset{\circ}{\overrightarrow{BC}} \cap \overset{\circ}{\overrightarrow{BE}} = \varnothing$.
e. $\overrightarrow{BC} \cup \overset{\circ}{\overrightarrow{BA}} = \overrightarrow{AC}$.
f. $\overrightarrow{BC} \cup \overrightarrow{BA} = \overleftrightarrow{AC}$.
g. $\overline{FE} \cap \overset{\circ}{\overrightarrow{BC}} = \varnothing$.
h. $\overrightarrow{AC} \cap \overline{FE} = B$.
i. $\overline{BD} \cup \overline{FE} = \{B\}$.
j. $\overrightarrow{AC} \cap \overline{FE} = \varnothing$.
k. $\overset{\circ}{\overrightarrow{BC}} \cup \overset{\circ}{\overrightarrow{BA}}$ is a convex set.
l. $\overline{CD}$ is a convex set.
m. $\overline{EB} \cup \overline{BC}$ is a convex set.
n. $\mathscr{L}_1$ is a convex set.

o. $\overline{AB} \cup \overline{CD}$ is a convex set.

p. $\overrightarrow{BA} \cup \overrightarrow{BE}$ is a convex set.

q. $\overrightarrow{BF}$ and $\overrightarrow{BE}$ are opposite rays.

r. $\overrightarrow{BF}$ and $\overrightarrow{BD}$ are opposite rays.

3. Let points, line segments, and lines which are labeled in the accompanying model be related *as suggested by the model.* For each of the following that are statements indicate whether the statement is true or false. In other cases complete the sentence to make a true statement.

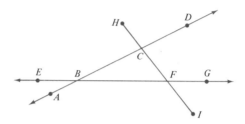

a. $B \in \overrightarrow{FE}$.

b. $\overrightarrow{EB} \cap \overrightarrow{BE} = ?$.

c. $\overline{BC} \subseteq \overset{\circ}{\overrightarrow{BC}}$.

d. $\overline{BC} \subseteq \overrightarrow{BC}$.

e. $\overline{HI} \cap (\overline{BC} \cup \overline{BF}) = ?$.

f. $\overline{BF} \cap (\overset{\circ}{\overrightarrow{BD}} \cup \overline{HI}) = ?$.

g. $\overleftrightarrow{AD} \cap \overrightarrow{BF} = ?$.

h. $\overline{BF} \cap (\overline{AC} \cup \overline{CI}) = ?$.

i. $\overline{BC} \cup \overline{CF} \cup \overline{BF}$ is a convex set.

j. $(\overline{EB} \cup \overline{BC}) \cap \overleftrightarrow{AD} = ?$.

k. $\overset{\circ}{\overrightarrow{BC}} \cap \overset{\circ}{\overrightarrow{BF}} = ?$.

l. $\overrightarrow{BE} \cup \overrightarrow{BF} = ?$.

4. If B is on a line $\mathscr{L}$ and separates $\mathscr{L}$ into half-lines such that P and Q are points on opposite half-lines, is $B \in \overline{PQ} \cap \mathscr{L}$? Draw a model that suggests the correctness of your answer.

5. Is the boundary point of two opposite half-lines an element of either half-line?

6. Draw models of plane figures which suggest each of the following:

a. $\overline{BC} \cup \overline{CD} \neq \overline{CD}$.

b. $\overrightarrow{AB} \subseteq \overleftrightarrow{CD}$.

c. $\overline{AB} \subseteq \overrightarrow{CD} \subseteq \overleftrightarrow{EF}$.

d. $\{C\} \subseteq \overset{\circ}{\overrightarrow{AB}} \subseteq \overrightarrow{DE}$.

e. $(\overline{BC} \cap \overline{PR}) \subseteq \overleftrightarrow{AS}$, $\overline{BC} \not\subseteq \overleftrightarrow{AS}$, and $\overline{PR} \not\subseteq \overleftrightarrow{AS}$.

f. $\overline{BC} \cap \{D\} \subseteq \overleftrightarrow{AS}$ and $\overline{BC} \nsubseteq \overleftrightarrow{AS}$.

g. $\{B,C\} \subseteq \overleftrightarrow{DS}$ and $B \notin \overset{\circ}{\overrightarrow{DS}}$.

h. $\{R,S\} \subseteq \overleftrightarrow{AB}$ and $\{R,S\} \nsubseteq \overset{\circ}{\overrightarrow{AB}}$.

3.4 SEPARATION OF PLANES, ANGLES

In many ways a line is related to a plane as a point is related to a line. For example, the set of all points on a line is infinite and the set of all lines in a plane is infinite. Another analogy is given in Postulate 3.2.

Postulate 3.2 (Plane separation postulate) For every plane α and for every line $\mathscr{L}$ in α, $\mathscr{L}$ separates the points in the plane α not on $\mathscr{L}$ into two convex sets such that if P is in one of the convex sets and Q is in the other, then $\overline{PQ} \cap \mathscr{L} \neq \varnothing$.

Definition 3.7 For every plane α and for every line $\mathscr{L}$ in α, the two convex subsets of α determined by $\mathscr{L}$ are called *half-planes*; the two half-planes determined by $\mathscr{L}$ are called *opposite* half-planes and $\mathscr{L}$ is called the *edge* or *boundary* of each half-plane.

We illustrate the ideas of Postulate 3.2 in Example 3.6.

EXAMPLE 3.6 Let plane α contain line $\mathscr{L}$ and points P and Q related as suggested in the accompanying model.

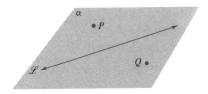

a. The sets of points in α not on $\mathscr{L}$ form two half-planes. That the half-planes (sets) are convex can be suggested by "picking any point X" on the model in the same half-plane as P and "drawing" the line segment PX. You will observe that every point on the line segment is in the same half-plane as P.

b. $\overline{PQ} \cap \mathscr{L} \neq \varnothing$.

For points and lines related as suggested in the model in Example 3.6 we will frequently refer to the half-plane which contains P as the P-side of $\mathcal{L}$ and will denote it by $\mathcal{L}:P$. Similarly $\mathcal{L}:Q$ denotes the half-plane which contains Q; that is, the Q-side of $\mathcal{L}$. If $\mathcal{L}$ is the line determined by points A and B (that is, $\overleftrightarrow{AB}$), then the P-side of line AB is denoted by $\overleftrightarrow{AB}:P$. Some illustrations are given in Example 3.7.

EXAMPLE 3.7 Let $\mathcal{L}$ be a line contained in plane α which contains points P and Q in opposite half-planes as suggested by the accompanying model. The following are true statements. (Recall! $\mathcal{L}:P$ and $\overleftrightarrow{AB}:P$ are names for the half-plane which contains P; that is, the P-side of $\overleftrightarrow{AB}$.)

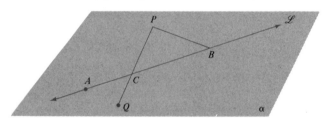

a. $\overleftrightarrow{AB}:P \cap \overleftrightarrow{AB}:Q = \varnothing.$ b. $\overleftrightarrow{AB}:P \cap \overleftrightarrow{AB} = \varnothing.$
c. $\overleftrightarrow{AB}:Q \cup \overleftrightarrow{AB}:P \neq \alpha.$ d. $\overline{PQ} \cap \overleftrightarrow{AB} = \{C\}.$
e. $\overline{PB} \not\subseteq \overleftrightarrow{AB}:P.$
f. For every point $X \in \overleftrightarrow{AB}:Q$, $\overline{QX} \subseteq \overleftrightarrow{AB}:Q.$

At this point we stop to summarize some remarks that were made somewhat incidentally in previous context: For every two points A and B,

1. the line which contains A and B is said to be determined by A and B and is denoted by $\overleftrightarrow{AB}$.

2. A and B also determine subsets of $\overleftrightarrow{AB}$ as follows:

    a. one line segment, $\overline{AB}$
    b. two rays, $\overrightarrow{AB}$ and $\overrightarrow{BA}$
    c. two half-lines, $\overset{\circ}{\overrightarrow{AB}}$ and $\overset{\circ}{\overrightarrow{BA}}$.

The concept of angle is important in geometry and is familiar to the reader. However, not every author gives the same definition for the concept of angle. We give a definition which best serves our needs and those of prospective elementary school teachers.

Definition 3.8 An *angle* is the union of two rays, not contained by the same line, with a common endpoint. The point which the two rays have in common is called the *vertex* of the angle. The two rays are called the *sides* of the angle.

Since an angle is the union of two rays and rays are sets of points, it follows that an *angle is a set of points*. An angle which is the union of $\overrightarrow{AB}$ and $\overrightarrow{AC}$ (see Fig. 3.6) we will denote by $\angle BAC$ or $\angle CAB$. That is, $\angle BAC = \overrightarrow{AB} \cup \overrightarrow{AC}$, or equivalently, $\angle CAB = \overrightarrow{AB} \cup \overrightarrow{AC}$. The reader might think about the question of what we mean when we write $\angle CAB = \angle BAC$, or more generally of what is meant when we write $\angle XYZ = \angle PQR$.

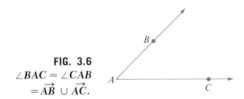

FIG. 3.6
$\angle BAC = \angle CAB$
$= \overrightarrow{AB} \cup \overrightarrow{AC}.$

Every three noncollinear points A, B, and C determine pairs of rays as follows: $\overrightarrow{BA}$ and $\overrightarrow{BC}$, $\overrightarrow{CB}$ and $\overrightarrow{CA}$, and $\overrightarrow{AB}$ and $\overrightarrow{AC}$; the unions of each of these pairs of rays are the following angles:$\angle ABC$, $\angle BCA$, and $\angle CAB$, respectively; therefore, we can think of three noncollinear points as determining three angles. The angle determined by three noncollinear points A, B, and C with vertex B, denoted by $\angle ABC$ or $\angle CBA$, is the set

$$\angle ABC = \overrightarrow{BA} \cup \overrightarrow{BC} = \{X \in S | X \in \overrightarrow{BA} \text{ or } X \in \overrightarrow{BC}\}.$$

When no ambiguity will result we denote $\angle ABC$ by $\angle B$. Some of these ideas are emphasized in Example 3.8.

EXAMPLE 3.8 Let $\overrightarrow{AB}$, $\overrightarrow{AC}$, $\overrightarrow{FD}$, and points A, . . . , F be related as suggested in the accompanying model. The following are true statements.

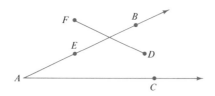

a. $\angle BAC = \overrightarrow{AB} \cup \overrightarrow{AC}$. b. $\angle BAC$ is *a set of points*.
c. $E \in \angle BAC$. d. $D \notin \angle BAC$. (Note this carefully!)
e. $F \notin \angle BAC$. f. $\overline{FD} \cap \angle BAC \neq \emptyset$.
g. $D \in \overleftrightarrow{AB}{:}C$. h. $D \in \overleftrightarrow{AC}{:}B$.

We call special attention to statement f. The model suggests that points D and F are related to $\angle BAC$ in much the same way that points which are in opposite half-planes determined by a line are related to the line. That is, $\angle BAC$ apparently separates the points in the plane, not on $\angle BAC$, into two disjoint subsets such that if P is in one set and Q in the other, then $\overleftrightarrow{PQ} \cap \angle BAC \neq \emptyset$. We refer to the two sets as the *interior* and *exterior* of the angle. The reader will probably agree that the model in Example 3.8 suggests that D is in the interior of $\angle BAC$ while F is in the exterior of $\angle BAC$.

This gives us an intuitive notion of interior and exterior of an angle; we have the necessary geometric concepts to make the idea precise. We refer to the model in Example 3.8 to abstract the essential ideas.

Since $C \notin \overleftrightarrow{AB}$, C is in a half-plane determined by $\overleftrightarrow{AB}$; that is, $C \in \overleftrightarrow{AB}{:}C$; similarly, B is in a half-plane determined by $\overleftrightarrow{AC}$; that is, $B \in \overleftrightarrow{AC}{:}B$. Now the point D (which we have intuitively agreed to say is in the interior of $\angle BAC$) is on the B-side of $\overleftrightarrow{AC}$ ($B \in \overleftrightarrow{AC}{:}B$) *and* on the C-side of $\overleftrightarrow{AB}$ ($C \in \overleftrightarrow{AB}{:}C$); that is, D is an element of the intersection set of the C-side of $\overleftrightarrow{AB}$ and the B-side of $\overleftrightarrow{AC}$ ($D \in \overleftrightarrow{AB}{:}C \cap \overleftrightarrow{AC}{:}B$). Thus, we are led to Definition 3.9.

Definition 3.9 For every A, B, and C which are noncollinear points, the interior of $\angle BAC$, denoted by $\mathrm{Int}(\angle BAC)$, is defined as: $\mathrm{Int}(\angle BAC) = \overleftrightarrow{AC}{:}B \cap \overleftrightarrow{AB}{:}C$.

Another look at the model in Example 3.8 will help to clarify Definition 3.9. Let us turn our attention to considerations which are relevant to describe the exterior of an angle, say $\angle BAC$. The model in Example 3.8 suggests that for a point P to be in the exterior of $\angle BAC$, denoted by $\mathrm{Ext}(\angle BAC)$, it would have to be true that:

1. $P \notin \angle BAC$, and
2. $P \notin \mathrm{Int}(\angle BAC)$.

We leave to the reader (Exercise 3 of Exercise Set 3.3) the problem of defining the exterior of $\angle BAC$ in terms of half-planes determined by lines AB and AC.

Exercise Set 3.3

1. Let $\angle ABC$, $\overleftrightarrow{DC}$, $\overline{EF}$, and points $A, \ldots, L$ be contained in plane α and be related as suggested by the accompanying model. Indicate whether the following are true:

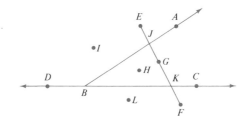

a. $\overrightarrow{BD}$ is a ray.
b. $\overrightarrow{BC}$ is a ray.
c. $\overrightarrow{BD} \cup \overrightarrow{BC}$ is an angle.
d. $H \in \mathrm{Int}(\angle ABC)$.
e. $\overrightarrow{BD} \cup \overrightarrow{BA}$ is an angle.
f. $\overline{EF} \cap \angle ABC = \overline{JK}$.
g. $\overline{EJ}$ is a line segment.
h. $\overline{EJ} \cup \overline{JA}$ is an angle.
i. $\overline{EF} \cap \angle ABC = \{E,F\}$.
j. $\overleftrightarrow{DC} \cap \angle ABC = \{B,C\}$.
k. $I \in \mathrm{Ext}(\angle ABC)$.
l. $I \in \mathrm{Int}(\angle DBA)$.

m. $\overrightarrow{BD} \cup \overrightarrow{BC} = \angle DBC.$

n. $\overrightarrow{JB} \cup \overrightarrow{BK} = \angle JBK.$

o. $\angle ABC \cap \overline{EF} = \{J,K\}.$

p. $\text{Int}(\angle ABC) \cap \overline{EF} = \overline{JK}.$

q. $\text{Ext}(\angle ABC) \cap \overrightarrow{BD} = \overrightarrow{BD}.$

r. $\text{Ext}(\angle ABC) \cap \text{Int}(\angle ABC) = \angle ABC.$

s. $\text{Ext}(\angle ABC) \cap \text{Int}(\angle ABC) = \emptyset.$

t. $\text{Ext}(\angle ABC) \cup \text{Int}(\angle ABC) = \alpha.$

u. $J \in \overleftrightarrow{DC}: J.$

v. $J \in \overleftrightarrow{DC}: A.$

w. $K \in \overleftrightarrow{DC}: J.$

x. $\overleftrightarrow{DC}: I \cap \overline{EF} = \overline{EK}.$

y. $\overleftrightarrow{DC}: F \cap \overleftrightarrow{DC}: G = \overleftrightarrow{DC}.$

z. $\overleftrightarrow{DC}: F \cup \overleftrightarrow{DC}: G = \text{plane } \alpha.$

A. $\overleftrightarrow{DC}: F \cup \overleftrightarrow{DC}: G = \text{plane } \alpha \setminus \overleftrightarrow{CD}.$

B. For every point $X \in \overleftrightarrow{BC}$, $X \in \overleftrightarrow{BC}: J.$

C. For every point $X \in \overleftrightarrow{BC}$, $X \notin \overleftrightarrow{BC}:L.$

D. For every point $X \in \text{Int}(\angle ABC)$, $X \notin \angle ABC.$

E. $\angle ABC \cup \text{Int}(\angle ABC) \cup \text{Ext}(\angle ABC) = \text{plane } \alpha.$

F. $\text{Int}(\angle ABC) \cap \overline{EF} = \overline{JK} \setminus \{J,K\}.$

2. Let $\overleftrightarrow{BC}, \overrightarrow{AD}, \overrightarrow{CD}$, and points $A, \ldots, I$ be related as suggested in the accompanying model. For each of the following, complete the sentence to make a true statement or answer the question.

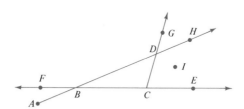

a. Name *all* of the angles suggested by the model.

b. $\overrightarrow{AH} \cap \angle FCG = ?.$

c. $\overrightarrow{CG} \cap \angle EBH = ?.$

d. B is the vertex of _____ .

e. $I \in \text{Int}(\ ) \cap \text{Int}(\ ).$

f. $\overline{BC} \cap \angle BCD = ?.$

g. Is $\overrightarrow{BF} \cup \overline{BA}$ an angle?.

h. $\overline{FC} \cap \angle FCG = ?$.
i. $C \in \angle$_____.
j. D is the vertex of _____.
k. What are the two sides of $\angle GDH$?
l. Is $\overline{CE}$ a side of $\angle GCE$?
m. Is $\overline{AB}$ the side of any angle?
n. Is D an element of the exterior of any angle suggested by the model? If yes, which one(s)?
o. Is G an element of the exterior of any angle suggested by the model? If yes, which one(s)?
p. $\overline{CD} \cap \text{Int}(\angle HBE) = ?$.
q. $\overline{CD} \cap \text{Ext}(\angle FBH) = ?$.
r. $\angle HBE \cap \angle GCE = ?$.
s. $\angle HBE \cap \angle GDH = ?$.
t. $\overrightarrow{AH} \cap \angle GCE = ?$.

3. Write a definition of the exterior of $\angle BAC$ using set builder notation. The accompanying model may be helpful.

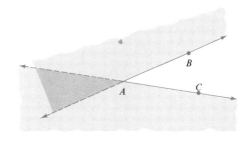

4. If P and Q are in opposite half-planes determined by a line $\mathcal{L}$, then:

a. How many points are contained in $\overline{PQ} \cap \mathcal{L}$?
b. Are any points of $\mathcal{L}$ in either half-plane?

5. Draw models of plane figures which suggest each of the following:

a. $\angle BAC \cap \overleftrightarrow{PQ} = \{B,C\}$.
b. $\text{Int}(\angle ABC) \cap \overleftrightarrow{PQ} = \overline{PQ} \setminus \{P,Q\}$.
c. $\text{Int}(\angle ABC) \cap \overrightarrow{BP} = \overrightarrow{BP}$.

d. $\angle ABC \cap \text{Int}(\angle EBF) = \angle ABC \setminus \{B\}$.

e. $\angle ABC \cup \angle CBD = \angle ABD \setminus \overset{\circ}{\overrightarrow{BC}}$.

f. $\angle ABC \cap \angle CBD = \overrightarrow{BC}$.

g. $\text{Int}(\angle ABC) \cap \text{Int}(\angle EBF) = \text{Int}(\angle EBC)$.

h. $\text{Int}(\angle ABC) \cup \text{Int}(\angle CBF) \cup \overset{\circ}{\overrightarrow{BC}} = \text{Int}(\angle ABF)$.

i. $P \notin \text{Int}(\angle CAB)$ and $P \notin \text{Ext}(\angle CAB)$.

4

---

distance, measure, congruence of line segments and angles

4.1 NUMBER SYSTEMS

The reader should be familiar with several number systems; probably each of the following: the system of whole numbers, the system of fractional numbers, the system of integers, the system of rational numbers, and the system of real numbers. Since properties of the system of real numbers will be used as we continue our study of geometry, we give a very brief summary of the properties of this number system.

4.1.1 The real number system

Every number system consists of a nonempty set on which is defined an equivalence relation (which is usually the equality relation) and two binary operations called "addition" and "multiplication." The system of real numbers consists of a nonempty set, which we denote by $\mathbb{R}$, and operations of addition and multiplication which enjoy the following properties.

COMMUTATIVE PROPERTIES

1. For all $a,b \in \mathbb{R}$, $a + b = b + a$.
2. For all $a,b \in \mathbb{R}$, $ab = ba$.

ASSOCIATIVE PROPERTIES

1. For all $a,b,c \in \mathbb{R}$, $(a + b) + c = a + (b + c)$.
2. For all $a,b,c \in \mathbb{R}$, $(ab)c = a(bc)$.

IDENTITIES

1. 0 is the unique real number such that for every $a \in \mathbb{R}$, $a + 0 = a$ and $0 + a = a$; 0 is the identity for addition.
2. $^{+}1$ is the unique real number such that for every $a \in \mathbb{R}$, $a \cdot {}^{+}1 = a$ and $^{+}1 \cdot a = a$; $^{+}1$ is the identity for multiplication.

INVERSES

1. For every $a \in \mathbb{R}$, there exists a unique real number $-a$ such that $a + (-a) = 0$ and $(-a) + a = 0$; $-a$ is the additive inverse of a.
2. For every $a \in \mathbb{R}$, $a \neq 0$, there exists a unique real number $a^{-1} \in \mathbb{R}$ such that $a \cdot a^{-1} = {}^{+}1$ and $a^{-1} \cdot a = {}^{+}1$; a^{-1} is the multiplicative inverse of a.   (Note a^{-1} is frequently denoted by $^{+}1/a$.)

DISTRIBUTIVE PROPERTIES

1. For all $a,b,c \in \mathbb{R}$, $a(b + c) = ab + ac$.
2. For all $a,b,c \in \mathbb{R}$, $(a + b)c = ac + bc$.

It is not difficult to prove that for all real numbers a and b, there exists a unique real number c such that $c + b = a$.   The real number c is the difference $a - b$ and this consideration leads to:

DEFINITION OF SUBTRACTION

For all $a,b \in \mathbb{R}$, $a - b = c$ if and only if $c + b = a$.

It is easily proved that the c in the above definition is the real number $a + (-b)$; that is,

For all $a,b \in \mathbb{R}$, $a - b = a + (-b)$.

Similar considerations about multiplication lead to:

DEFINITION OF DIVISION

For all $a,b \in \mathbb{R}$, $b \neq 0$, $a \div b = c$ if and only if $cb = a$.

It is easily proved that the c in the above definition is the real number $a \cdot b^{-1}$; that is,

For all $a,b \in \mathbb{R}$, $b \neq 0$, $a \div b = a \cdot b^{-1}$.

Certain properties of addition and equality enable us to write equations which are equivalent to given equations. They are:

THE ADDITION–EQUATION TRANSFORMATION PRINCIPLE

For all $a,b,c \in \mathbb{R}$, $a = b$ if and only if $a + c = b + c$.

THE MULTIPLICATION–EQUATION TRANSFORMATION PRINCIPLE

For all $a,b,c \in \mathbb{R}$, $c \neq 0$, $a = b$ if and only if $ac = bc$.

We assume that the reader is familiar with the concepts of positive and negative real numbers. This enables us to define the order relation of less than ($<$) on $\mathbb{R}$ as follows:

DEFINITION OF LESS THAN

For all $a,b \in \mathbb{R}$, $a < b$ if and only if there exists a positive real number c such that $a + c = b$.

The closely associated relation of greater than is as follows:

DEFINITION OF GREATER THAN

For all $a,b \in \mathbb{R}$, $a > b$ if and only if there exists a positive real number c such that $a = b + c$.

It is an important property that for all real numbers a and b, a and b are related by one of the three relations: equality, less than, or greater than. The formal statement of this property is called:

THE TRICHOTOMY

For all $a,b \in \mathbb{R}$, exactly one of the following is true:

$a < b$, $a = b$, $a > b$.

Properties of addition, multiplication, and less than give the following transformation principles:

THE ADDITION–INEQUALITY TRANSFORMATION PRINCIPLE

For all $a,b,c \in \mathbb{R}$, $a < b$ if and only if $a + c < b + c$.

THE MULTIPLICATION–INEQUALITY TRANSFORMATION PRINCIPLES

1. For all $a,b,c \in \mathbb{R}$, $c > 0$, $a < b$ if and only if $ac < bc$.
2. For all $a,b,c \in \mathbb{R}$, $c < 0$, $a < b$ if and only if $ac > bc$.
3. For all $a,b,c \in \mathbb{R}$, $c > 0$, $a > b$ if and only if $ac > bc$.
4. For all $a,b,c \in \mathbb{R}$, $c < 0$, $a > b$ if and only if $ac < bc$.

A property of the system of real numbers which distinguishes it from other number systems is:

THE COMPLETENESS PROPERTY

There exists a one-to-one correspondence between the set of real numbers and the points of a line.

An important function on the set $\mathbb{R}$ which we will use is:

THE ABSOLUTE VALUE FUNCTION

For every $a \in \mathbb{R}$, $|a| = a$ if $a \geqslant 0$ and $|a| = -a$ if $a < 0$.

Exercise set 4.1

1. For each of the following statements, name the property of real numbers which justifies it.   For example, "For $x \in \mathbb{R}$, $x + {}^+3 < {}^+5 + {}^+3$ if and only if $x < {}^+5$" is true because of the addition–inequality transformation principle.

    a. ${}^+3 \cdot {}^-4 = {}^-4 \cdot {}^+3$.
    b. ${}^-3({}^-5 + {}^+4) = {}^-3 \cdot {}^-5 + {}^-3 \cdot {}^+4$.
    c. ${}^-317 \cdot {}^+21 + {}^-317 \cdot {}^-1 = {}^-317({}^+21 + {}^-1)$.
    d. $({}^-3 + {}^+5) + {}^-12 = {}^-3 + ({}^+5 + {}^-12)$.
    e. ${}^+5 + 0 = 0 + {}^+5$.
    f. $\sqrt{{}^+3} \cdot {}^+1 = \sqrt{{}^+3}$.
    g.   For all $x,y \in \mathbb{R}$, if $x \cdot {}^-5 < y \cdot {}^-5$, then $x > y$.

h. For all $x,y \in \mathbb{R}$, if $\dfrac{x+y}{^{+}2} < {}^{+}5$, then $\dfrac{x+y}{^{+}2} + {}^{-}2 < {}^{+}5 + {}^{-}2$.

i. $\sqrt{^{+}7} \cdot \dfrac{1}{\sqrt{^{+}7}} = {}^{+}1$.

j. For every $x \in \mathbb{R}$, $\sqrt{^{+}5} < x$ if and only if ${}^{+}3 \cdot \sqrt{^{+}5} < {}^{+}3 \cdot x$.

k. For every $x \in \mathbb{R}$, $\sqrt{^{+}13} > x$ if and only if ${}^{-}1 \cdot \sqrt{^{+}13} < {}^{-}1 \cdot x$.

l. ${}^{+}5 + \sqrt{^{+}21} > {}^{+}5 + \sqrt{^{+}5}$; therefore, $\sqrt{^{+}21} > \sqrt{^{+}5}$.

m. $\dfrac{^{+}5}{^{+}7} \cdot \dfrac{^{+}7}{^{+}5} = {}^{+}1$.

n. ${}^{-}5 + {}^{+}5 = 0$.

o. For every $m \in \mathbb{R}$, if ${}^{-}3m = {}^{+}12$, then $m = {}^{-}4$.

2. For all $x,y \in \mathbb{R}$, $x \leqslant y$ means that $x < y$ or $x = y$. For all $x,y,z \in \mathbb{R}$, $x < y < z$ means that $x < y$ and $y < z$. Write each of the following as a conjunction or disjunction.

    a. ${}^{+}5 \leqslant {}^{+}5$.   **b.** ${}^{-}3 < 0 < {}^{+}7$.

3. Prove the following:

    a. For all $a,b \in \mathbb{R}$, $-(a+b) = (-a) + (-b)$.

    b. For all $a,b \in \mathbb{R}$, $-(a-b) = b - a$.

    c. For all $a,b \in \mathbb{R}$, $|a-b| = a - b$ if $a \geqslant b$, and $|a-b| = b - a$ if $a < b$.

4. Determine each of the following:

    a. $|{}^{+}5|$ **b.** $|{}^{-}31|$ **c.** $|{}^{-}7 + {}^{-}4|$

    d. $|{}^{+}7 - {}^{-}3|$ **e.** $|{}^{-}8 + {}^{+}2|$ **f.** $|{}^{-}3 - {}^{+}7|$

4.2 DISTANCE AND ITS MEASURE

We shall accept the concept of *distance* or the *distance from one point to another point* as an undefined concept in our development of geometry. We turn our attention to a discussion of the measure of distance which we will find to be basic to all ideas about measure in geometry.

Since space is an infinite set of points we are at liberty to choose a set of *two* points, $\{O,I\}$; we shall call such a set a *unit pair*. The real number ${}^{+}1$, which we assign to the ordered unit pair (O,I) is the measure of the distance from O to I. Then, for every A and B which are points in space, the measure of the distance from A to

B is determined by "comparing" the distance from A to B with the distance from O to I; exactly how this "comparison" is accomplished we specify in a sequence of postulates. First of all we guarantee that for every A and B which are points, the distance from A to B has a measure with respect to a chosen unit pair.

Postulate 4.1   For every unit pair $U = \{O,I\}$ there exists a function m_U, called a *distance function*, which assigns to the unit pair the real number $^+1$ and assigns to every other two points a unique nonnegative real number.

Definition 4.1   For every $(A,B) \in S \times S$, the *measure* of the distance from A to B, relative to the unit pair $U = \{O,I\}$, is the nonnegative real number which the distance function m_U assigns to (A,B).

We will denote the measure of the distance from A to B relative to $U = \{O,I\}$ by $m_U(A,B)$ which we shall read as the "measure of the distance from A to B with respect to U" or "the U-measure of the distance from A to B."

We call the reader's attention to the fact that every unit pair determines a distance function; therefore, a measure of the distance from a point A to a point B depends on the chosen unit pair. That is, for most choices of unit pairs $U_1 = \{O_1,I_1\}$ and $U_2 = \{O_2,I_2\}$, and for all points $A,B \in S$, $m_{U_1}(A,B) \neq m_{U_2}(A,B)$.

We make assumptions about the distance functions as given in Postulate 4.2.

Postulate 4.2 (Distance function postulate)   For every unit pair $U = \{O,I\}$:

D1   For all $A,B \in S$, $m_U(A,B) = 0$ if and only if $A = B$.

D2   For all $A,B \in S$, $m_U(A,B) > 0$ if and only if $A \neq B$.

D3   For all $A,B \in S$, $m_U(A,B) = m_U(B,A)$.

D4   For all $A,B,C \in S$, $m_U(A,B) + m_U(B,C) \geq m_U(A,C)$.

Property D4 above is called the *triangle inequality*; the models in Fig. 4.1 suggest the reason for this name.

Because of property D3 it is usually unnecessary to distinguish

FIG. 4.1

Model to suggest triangle inequality.

(a) $m_U(A,B) + m_U(B,C)$
$> m_U(A,C)$.

(b) $m_U(D,E) + m_U(E,F)$
$= m_U(D,F)$.

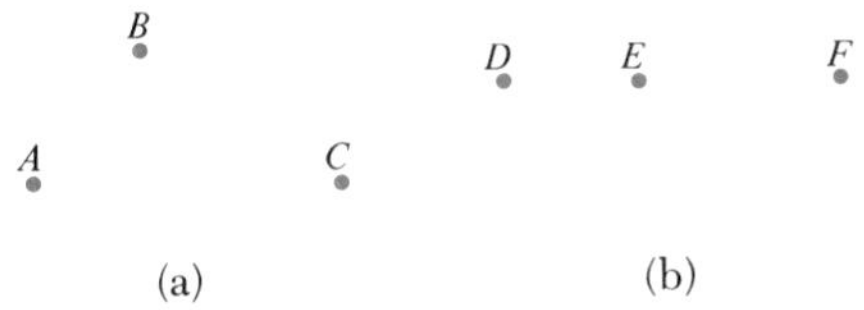

(a)

(b)

between the distance from A to B and the distance from B to A; in such cases we will simply say the distance *between* A and B and denote the measure with respect to $U = \{O,I\}$ by $m_U(A,B)$ or $m_U(B,A)$.

Now, Postulate 4.1 guarantees that to every A and B in space there is assigned, relative to a given unit pair, a unique nonnegative real number—Definition 4.1 states that this is a measure of the distance between them—however, as yet we do not know how to determine the real number which is assigned to a pair of points. That is, as yet we do not know how to "compare" the distance between two points in space to the distance between the unit pair. To do this we appeal to the geometric analog of a ruler; namely, a coordinate system on a line.

Postulate 4.3 For every unit pair $U = \{O,I\}$, there exists a one-to-one correspondence between the line $\mathscr{L}$ which contains O and I, and the set of real numbers such that the coordinates of O and I are 0 and $^+1$, respectively; and for every A and B on $\mathscr{L}$ with coordinates a and b, $m_U(A,B) = |b - a|$.

The particular one-to-one correspondence between the set of real numbers and a line determined by a given unit pair is called the *coordinate system of the line determined by the unit pair.* We illustrate Postulate 4.3 in Example 4.1.

EXAMPLE 4.1

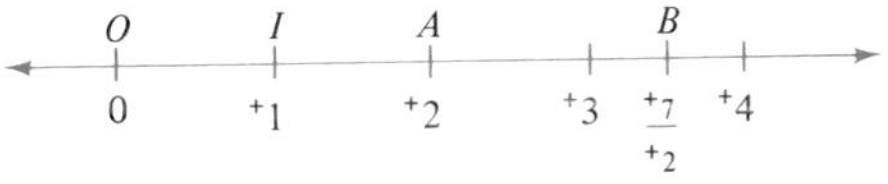

$$m_U(A,B) = \left| \frac{^+7}{^+2} - {^+2} \right|$$

$$= \left| \frac{^+3}{^+2} \right|$$

$$= \frac{^+3}{^+2}.$$

$$\therefore m_U(A,B) = \frac{^+3}{^+2}.$$

4   distance, measure, congruence of line segments and angles

Now that we have the measure of distance between two points related to a coordinate system on a line, there remains only one question yet to be answered to be sure we can determine the distance between *every* pair of points. We can, provided there is always a line with a coordinate system such that the two points have coordinates in that coordinate system. This is guaranteed in Postulate 4.4.

Postulate 4.4 (The ruler placement postulate)   For every two points A and B, there exists a coordinate system on the line determined by A and B relative to some unit pair such that the coordinate of A is zero and the coordinate of B is a positive real number.

Many geometric relationships which we develop in this chapter depend ultimately on the measure of the distance between two points. We now have the somewhat distressing situation that the distance between two distinct points A and B has infinitely many measures, one corresponding to every choice of a unit pair. This raises the serious question about whether geometric relationships also depend on the choice of a unit pair. To answer this question we have Postulate 4.5.

Postulate 4.5   If $U_1 = \{O_1, I_1\}$ and $U_2 = \{O_2, I_2\}$ are two unit pairs which determine two coordinate systems on a line $\mathscr{L}$, then for all points P and Q on $\mathscr{L}$, $m_{U_1}(P, Q) = m_{U_1}(O_2, I_2) \times m_{U_2}(P, Q)$.

We illustrate Postulate 4.5 in Example 4.2.

EXAMPLE 4.2   Let $U_1 = \{O_1, I_1\}$ and $U_2 = \{O_2, I_2\}$ be two unit pairs which determine two coordinate systems on a line $\mathscr{L}$ as suggested in the accompanying model. (The coordinates of points on the line with respect to $\{O_1, I_1\}$ are indicated "above $\mathscr{L}$" and with respect to $\{O_2, I_2\}$ "below $\mathscr{L}$.")

$$m_{U_1}(P, Q) = |{}^+6 - {}^+3|$$
$$= {}^+3.$$

$$m_{U_2}(P,Q) = |{}^+4 - {}^+2|$$
$$= {}^+2.$$

$$m_{U_1}(O_2,I_2) = \left|\frac{{}^+3}{{}^+2} - 0\right|$$
$$= \frac{{}^+3}{{}^+2}.$$

Now, ${}^+3 = \dfrac{{}^+3}{{}^+2} \times {}^+2$; therefore,

$$m_{U_1}(P,Q) = m_{U_1}(O_2,I_2) \times m_{U_2}(P,Q).$$

A result which is immediate from Postulate 4.5 is given in Theorem 4.1.

Theorem 4.1 If $U_1 = \{O_1,I_1\}$ and $U_2 = \{O_2,I_2\}$ are two distinct unit pairs which determine two coordinate systems on a line $\mathscr{L}$ and $m_{U_1}(O_2,I_2) = {}^+1$, then for all points P and Q on $\mathscr{L}$,

$$m_{U_1}(P,Q) = m_{U_2}(P,Q).$$

The proof is left as an exercise for the reader.

What Postulate 4.5 tells us is that we can always "convert" the measure of the distance between two points determined by one coordinate system to the measure determined by another coordinate system (or equivalently, a measure determined with respect to one unit pair can be "converted" to a measure with respect to another unit pair); this fact guarantees that geometric relationships stated in terms of measure are independent of the choice of the unit pair. For this reason, hereafter, when we refer to *the* measure of the distance between two points we will not make reference to a unit except where emphasis is required; or what is probably more accurate, we will assume that we have a fixed unit pair.

Since all statements involving the measure of distance will be made relative to a fixed unit pair we simplify our notation as follows: For all $A,B \in S$, we denote the measure of the distance from A to B by AB, the measure of the distance from B to A by BA. Usually we are not concerned whether the measure is of the distance from A to B or from B to A; in this case we refer to the measure of the distance *between* A and B, and denote it by AB or BA. The distance function postulate is then as follows:

4 distance, measure, congruence of line segments and angles

D1   For all $A,B \in S$, $AB = 0$ if and only if $A = B$.

D2   For all $A,B \in S$, $AB > 0$ if and only if $A \neq B$.

D3   For all $A,B \in S$, $AB = BA$.

D4   For all $A,B,C \in S$, $AB + BC \geq AC$.

A remark   The use of a coordinate system to determine the measure of the distance between two points is the geometric abstraction of using a ruler to determine the measure of the distance between two physical objects. Postulate 4.5 is the geometric abstraction of converting, for example, a "measure" obtained by using an inch-ruler to a "measure" determined by using a foot-ruler.

Recall that in Chapter 3 we proceeded with a naive notion of betweenness; we are now in a position to give a precise definition of the betweenness relation on S.

Definition 4.2   For all $A,B,C \in S$, B is *between* A and C if and only if

1. A, B, and C are collinear, and
2. $AB + BC = AC$.

Exercise set 4.2

1. Consider points A, B, C, D, and O and the coordinate system on line $\mathcal{L}$ as suggested in the accompanying model. Determine:

    a. $AB = |\underline{\hspace{1cm}} - \underline{\hspace{1cm}}| = \underline{\hspace{1cm}}$.
    b. $BC = |\underline{\hspace{1cm}} - \underline{\hspace{1cm}}| = \underline{\hspace{1cm}}$.
    c. $OC = ?$.   d. $DB = ?$.   e. $BD = ?$.
    f. $AD = ?$.   g. $AO = ?$.   h. $DO = ?$.

2. Prove: For all $A,B,C \in S$, A–B–C if and only if C–B–A.

3. Prove: For all $A,B \in S$, $\overline{AB} = \overline{BA}$. (*Hint:* Prove $\overline{AB} \subseteq \overline{BA}$ and $\overline{BA} \subseteq \overline{AB}$.)

4. Let $U_1 = \{O_1, I_1\}$ and $U_2 = \{O_2, I_2\}$ be two unit pairs which determine two coordinate systems on a line $\mathscr{L}$ as suggested by the accompanying model. (Coordinates with respect to U_1 are given "above $\mathscr{L}$" and coordinates with respect to U_2 are given "below $\mathscr{L}$.")

a. (1) $m_{U_1}(P,Q) = |\underline{\hspace{1cm}} - \underline{\hspace{1cm}}| = \underline{\hspace{1cm}}$.
(2) $m_{U_2}(P,Q) = ?$.   (3) $m_{U_1}(O_2,I_2) = ?$.
(4) Is $m_{U_1}(P,Q) = m_{U_1}(O_2,I_2) \times m_{U_2}(P,Q)$? Explain your answer.

b. (1) $m_{U_1}(P,M) = ?$.   (2) $m_{U_2}(P,M) = ?$.
(3) $m_{U_1}(P,M) = \underline{\hspace{1cm}} \times m_{U_2}(P,M)$.

c. $m_{U_1}(Q,M) = \underline{\hspace{1cm}} \times m_{U_2}(Q,M)$.

5. Let $U_1 = \{O_1, I_1\}$ and $U_2 = \{O_2, I_2\}$ be two unit pairs which determine two coordinate systems on a line $\mathscr{L}$ as suggested by the accompanying model. (Coordinates with respect to U_1 are given "above $\mathscr{L}$" and coordinates with respect to U_2 are given "below $\mathscr{L}$.")

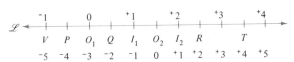

a. (1) $m_{U_1}(T,V) = ?$.   (2) $m_{U_2}(T,V) = ?$.
(3) $m_{U_1}(O_2,I_2) = ?$.   (4) $m_{U_2}(O_1,I_1) = ?$.
(5) Is $m_{U_1}(T,V) = m_{U_1}(O_2,I_2) \times m_{U_2}(T,V)$? Explain your answer.

b. Answer questions (1), (2), and (5) above for:
(1) (R,P) (2) (T,O_1) (3) (O_1,O_2)
(4) (I_2,O_1) (5) (I_2,V) (6) (P,Q).

4.3 MEASURE OF LINE SEGMENTS, CONGRUENT LINE SEGMENTS

The concept of congruence is basic to classification, comparison, and measurement of geometric figures. The intuitive notion of congruence is that of "same size" and "same shape." However, "size" and "shape" refer to physical notions, therefore, we need to abstract these to geometric concepts. We first look at the concept of congruence and measure as it applies to line segments.

In a previous chapter we proved that every line can be characterized as being straight in the sense that through two points there is exactly one line. Thus the word "straight" describes what we mean by the "shape of a line" and hence also the "shape of a line segment." The geometric abstraction of the "size" of a line segment is given in Definition 4.3.

Definition 4.3 For every line segment AB, the measure of $\overline{AB}$, denoted by $m(\overline{AB})$, is the *measure* of the distance between A and B.

A remark A word that is often used in connection with line segments is "length." The *length* of a line segment is the measure (a real number) together with the unit used to determine the measure. For example, if the inch-measure of the model of a line segment is $^+3$, then we say the length of the line segment is $^+3$ inches. We will have more to say about length in the next section where we deal with the process, practice, or application of the theory of measure.

Now with the geometric abstractions of size and shape defined we are in a position to consider the relation of congruence on line segments.

Definition 4.4 For all line segments AB and CD, $\overline{AB}$ is *congruent* to $\overline{CD}$ if and only if $m(\overline{AB}) = m(\overline{CD})$. If $\overline{AB}$ is congruent to $\overline{CD}$, we write $\overline{AB} \cong \overline{CD}$.

We state Theorem 4.2, but leave the proof as an exercise.

Theorem 4.2 The relation $\cong$ on line segments has the following properties:

a. For every line segment AB, $\overline{AB} \cong \overline{AB}$.
b. For all line segments AB and CD, if $\overline{AB} \cong \overline{CD}$, then $\overline{CD} \cong \overline{AB}$.
c. For all line segments AB, CD, and EF, if $\overline{AB} \cong \overline{CD}$ and $\overline{CD} \cong \overline{EF}$, then $\overline{AB} \cong \overline{EF}$.

That is, $\cong$ is an equivalence relation on the set of all line segments.

Definition 4.5 For every line segment AB, a point $X \in \overline{AB}$ is the *midpoint* of $\overline{AB}$ if and only if $\overline{AX} \cong \overline{XB}$.

The midpoint of a line segment is said to *bisect* the line segment; similarly a line, ray, half-line, or line segment which intersects a line segment only in its midpoint is said to bisect the line segment.

Exercise set 4.3

1. Consider the points on the real number line with coordinates as indicated in the accompanying figure.

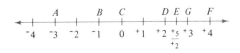

a. $m(\overline{FD}) = |\underline{\hspace{1cm}} - \underline{\hspace{1cm}}| = \underline{\hspace{1cm}}.$
b. $m(\overline{DF}) = |\underline{\hspace{1cm}} - \underline{\hspace{1cm}}| = \underline{\hspace{1cm}}.$
c. $m(\overline{DC}) = |\underline{\hspace{1cm}} - \underline{\hspace{1cm}}| = \underline{\hspace{1cm}}.$
d. $m(\overline{DE}) = ?.$ e. $m(\overline{BC}) = ?.$ f. $m(\overline{CA}) = ?.$

2. Refer to the number line in Exercise 1. Indicate whether the following are true or false. Explain your answer.

a. $\overline{AB} \cong \overline{CD}.$
b. C is the midpoint of $\overline{AE}.$
c. D is the midpoint of $\overline{BF}.$
d. C bisects $\overline{AE}.$
e. $m(\overline{DE}) = m(\overline{EG}).$
f. $\overline{DE} \cong \overline{EG}.$
g. $\overline{DE} \cong \overline{ED}.$
h. $m(\overline{DF}) = m(\overline{FD}).$
i. $m(\overline{BC}) = \frac{+1}{+3} \times m(\overline{AC}).$
j. $\overline{BC} < \overline{BD}.$
k. $\overline{BC} = \overline{DE}.$
l. $BC = DE.$
m. $BC = \frac{+1}{+3} \times AC.$
n. $BC < BD.$
o. $m(\overline{BC}) < m(\overline{BD}).$
p. $AB + BC = AC.$
q. $\overline{AB} + \overline{BC} = \overline{AC}.$
r. $m(\overline{AB}) + m(\overline{BC}) = m(\overline{AC}).$
s. $AC + AE > AD.$
t. $m(\overline{AC}) + m(\overline{AE}) > m(\overline{AD}).$

3. Prove Theorem 4.2.

4. Prove: For all $A,B \in S$, $\overline{AB} \cong \overline{BA}$.

5. Prove: For all $A,B,C,D,E,F \in S$, if A–B–C, D–E–F, $\overline{AB} \cong \overline{DE}$, and $\overline{BC} \cong \overline{EF}$, then $\overline{AC} \cong \overline{DF}$.

6. Prove: For all $A,B,C,D,E,F \in S$, if A–B–C, D–E–F, $\overline{AB} \cong \overline{DE}$, and $\overline{AC} \cong \overline{DF}$, then $\overline{BC} \cong \overline{EF}$.

7. Draw a model to suggest that the following statement is false: For all $\overline{AB}$, $\overline{CD}$, $\overline{EF}$, if $\overline{AB} \cong \overline{CD}$, then $\overline{AB} \cup \overline{EF} \cong \overline{CD} \cup \overline{EF}$.

8. Draw models to suggest whether each of the following statements is true or false:

    a. For all $\overline{AB}$, $\overline{CD}$, $\overline{EF}$, if $\overline{AB} \cup \overline{EF} \cong \overline{CD} \cup \overline{EF}$, then $\overline{AB} \cong \overline{CD}$.
    b. For all $\overline{AX}$, $\overline{XB}$, if $\overline{AX} \cong \overline{XB}$, then X is the midpoint of $\overline{AB}$.
    c. For all $\overline{AB}$, $\overline{CD}$, $\overline{EF}$, if $m(\overline{AB}) + m(\overline{CD}) = m(\overline{EF})$, then $\overline{AB} \cup \overline{CD} \cong \overline{EF}$.
    d. For all $\overline{AB}$, $\overline{CD}$, $\overline{EF}$, if $m(\overline{AB}) + m(\overline{EF}) = m(\overline{CD}) + m(\overline{EF})$, then $\overline{AB} \cong \overline{CD}$.
    e. For all $\overline{AB}$, $\overline{CD}$, $\overline{EF}$, if $\overline{AB} \cong \overline{CD}$, then $m(\overline{AB}) + m(\overline{EF}) = m(\overline{CD}) + m(\overline{EF})$.

4.4 APPLICATION OF THE THEORY OF MEASURE

The theory which we have presented concerning geometric measure is related to and is an abstraction from "everyday" concepts related to the measure of physical objects and models of line segments.   One significant difference between the theoretical notions and the application of these is that theoretically we can determine *exactly* the measure of a line segment while we can only *approximate* the measure of models of line segments.

To suggest why this is true and to see how the theory of measure is related to the practice of measuring physical models of line segments, some thoughts about a ruler and the use of a ruler may be appropriate.

First of all, a ruler is developed in a manner that is analogous to the way a coordinate system is defined on a line; that is, on a "straight edge" some "point" is chosen as the "zero point" (the origin) and is labeled with a "0." Then, a second "point" is chosen and labeled with a "1" and this pair of points determines the "unit of measure." After this, successive "points" which are two, three, . . . "units" from the "zero point" are "laid off" and labeled with a "2," "3," and so on. (Frequently, because of social reasons a unit called an "inch" is chosen and an inch-ruler similar to the one shown in the accompanying figure might be the result.)

Now to use this ruler to approximate the measure of the accompanying model of line segment AB, we would pick up our ruler

and lay it down along the model of $\overline{AB}$ in such a way that the "zero point" is at A as suggested in the accompanying model.

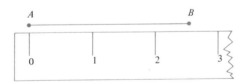

(*Note:* Think carefully and identify the axioms which describe the geometric abstraction of the ruler manipulations described above.) We observe that the "point B" does not correspond exactly with any of the labeled "points" on this ruler; since B is closer to the "point" labeled 3, we report that the measure of the model line segment is approximately +3 and write $m(\overline{AB}) \doteq {}^{+}3$. ["$\doteq$" is read, "is approximately equal to."] If we would bisect each of the model unit line segments as suggested in the accompanying model, we could report that the measure of the model of $\overline{AB}$ is approximately ${}^{+}2\frac{1}{2}$.

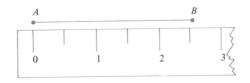

4 distance, measure, congruence of line segments and angles

The first approximation of the measure of the model of $\overline{AB}$ was given correct to the nearest whole unit; the second approximation was given correct to the nearest half-unit. That is,

$m(\overline{AB}) \doteq {}^+3$ correct to the nearest whole unit;
$m(\overline{AB}) \doteq {}^+2\frac{1}{2}$ correct to the nearest half-unit.

If we subdivided each of the model unit line segments into 10 "congruent" model line segments, then we could approximate the measure of the model of $\overline{AB}$ correct to the nearest $\frac{1}{10}$ of a unit. Suppose our approximation indicates that $m(\overline{AB}) \doteq {}^+2.6$; moreover, suppose also that after further subdivision of each unit line segment into 100 congruent parts we determined that $m(\overline{AB}) \doteq {}^+2.64$.

We would probably agree that each of the successive approximations given for the measure of the model of $\overline{AB}$ is "better" than the previous one. Let us look more carefully at the sense in which each one is successively "better." The accompanying model suggests that we would report the measure (correct to the nearest whole unit) as $^+3$ for every line segment whose true measure is greater than or equal to $^+2.5$ and less than $^+3.5$.

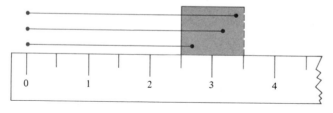

That is,

$m(\overline{AB}) \doteq {}^+3$

means that

$^+2.5 \leqslant m(\overline{AB}) < {}^+3.5.$

Similarly,

$m(\overline{AB}) \doteq {}^+2\frac{1}{2}$

means that

$^+2\frac{1}{4} \leqslant m(\overline{AB}) < {}^+2\frac{3}{4}.$

Likewise,

$m(\overline{AB}) \doteq {}^+2.6$

means that

${}^+2.55 \leqslant m(\overline{AB}) < {}^+2.65,$

and

$m(\overline{AB}) \doteq {}^+2.64$

means that

${}^+2.635 \leqslant m(\overline{AB}) < {}^+2.645.$

To summarize: A report that the approximation of the measure of model line segments is ${}^+3$ (which also indicates that the approximation is correct to the nearest whole unit) can apply to model line segments whose true measures could differ from each other by one whole unit. That is, there is an "interval of uncertainty" associated with this approximation which is one unit.

An approximation of the measure of the model of $\overline{AB}$ given as ${}^+2\frac{1}{2}$ (indicating an approximation correct to the nearest half-unit) could apply to model line segments whose true measures differ from each other by as much as one-half of a unit. That is, the interval of uncertainty for this approximation is one-half of a unit. Similarly, an approximate measure of ${}^+2.6$ (which indicates an approximation correct to the nearest $\frac{1}{10}$ of a unit) can apply to model line segments whose true measures could differ from each other by $\frac{1}{10}$ of a unit. What is the interval of uncertainty for this approximation? We leave to the reader the problem of describing similarly the approximation ${}^+2.64$.

The measure of the interval of uncertainty associated with an approximation of a measure is called the *precision* of the measure. That is, one word to indicate that each of the approximations given in our discussion above is successively "better" is that each is successively more *precise*. From our illustrations above it is apparent that smaller "subdivisions of the unit" result in more precise approximations; or equivalently, a smaller unit of measure results in more precise approximations.

The greatest possible difference between a true measure and an approximation of this measure is called the *greatest possible error* (g.p.e.). This difference is one-half of the precision of the approximation.

We illustrate further the ideas of precision and g.p.e in Example 4.3.

EXAMPLE 4.3

a. If $m(AB) \doteq {}^{+}5$, then:
   (1) ${}^{+}4.5 \leqslant m(\overline{AB}) < {}^{+}5.5$,
   (2)   the precision of the approximation is ${}^{+}5.5 - {}^{+}4.5$, or ${}^{+}1$,
   (3)   the g.p.e. of the approximation is $\frac{{}^{+}1}{{}^{+}2} \times {}^{+}1$, or ${}^{+}0.5$.

b. If $m(\overline{AB}) \doteq {}^{+}2.67$, then:
   (1) ${}^{+}2.665 \leqslant m(\overline{AB}) < {}^{+}2.675$,
   (2)   the precision of the approximation is ${}^{+}2.675 - {}^{+}2.665$ or ${}^{+}0.01$,
   (3)   the g.p.e. of the approximation is $\frac{{}^{+}1}{{}^{+}2} \times {}^{+}0.01$ or ${}^{+}0.005$.

c. If $m(\overline{AB}) \doteq {}^{+}2\frac{1}{8}$, then:
   (1) ${}^{+}2\frac{1}{16} \leqslant m(\overline{AB}) < {}^{+}2\frac{3}{16}$,
   (2)   the precision of the approximation is

$$ {}^{+}2\frac{3}{16} - {}^{+}2\frac{1}{16} = \frac{{}^{+}2}{{}^{+}16} \text{ or } \frac{{}^{+}1}{{}^{+}8}, $$

   (3)   the g.p.e. of the approximation is $\frac{{}^{+}1}{{}^{+}2} \times \frac{{}^{+}1}{{}^{+}8}$ or $\frac{{}^{+}1}{{}^{+}16}$.

So far we have considered the actual amount of error that may result in approximating a measure. Sometimes the actual error that may occur in approximating a measure is not as important as the "relative size" of the error compared to the approximation itself. For example, an approximation of the measure of the edge of a table top correct to the nearest inch would have a g.p.e. of $\frac{{}^{+}1}{{}^{+}2}$. Also an approximation of the measure of the distance from a "point" in Chicago to a "point" in Pittsburgh correct to the nearest inch would have the same g.p.e. However, one would certainly be more impressed with the approximation of the measure of the distance from Chicago to Pittsburgh. Why? The reason is easily seen by comparing the ratios of the g.p.e. of each approximate measure to the corresponding approximate measure. This ratio— g.p.e. of an approximate measure to the corresponding approximate measure—is called the *relative error* (r.e.) of the approximation. Greatest possible error is associated with the precision of an approximation; relative error is associated with the *accuracy* of the approximation. The smaller the relative error the greater the accuracy. These ideas are illustrated in Example 4.4.

EXAMPLE 4.4

a. If $m(\overline{QD}) \doteq {}^+3.8$, then:
   (1) ${}^+3.75 \leqslant m(\overline{QD}) < {}^+3.85$,
   (2) the precision of the approximation is ${}^+0.1$,
   (3) the g.p.e. of the approximation is ${}^+0.05$,
   (4) the r.e. of the approximation is $\dfrac{{}^+0.05}{{}^+3.8}$.

b. If $m(\overline{PR}) \doteq {}^+197.6$, then:
   (1) ${}^+197.55 \leqslant m(\overline{PR}) < {}^+197.65$,
   (2) the precision of the approximation is ${}^+0.1$,
   (3) the g.p.e. of the approximation is ${}^+0.05$,
   (4) the r.e. of the approximation is $\dfrac{{}^+0.05}{{}^+197.6}$.

c. If $m(\overline{TU}) \doteq {}^+2\frac{5}{32}$, then:
   (1) ${}^+2\frac{9}{64} \leqslant m(\overline{TU}) < {}^+2\frac{11}{64}$,
   (2) the precision of the approximation is $\dfrac{{}^+2}{{}^+64}$ or $\dfrac{{}^+1}{{}^+32}$,
   (3) the g.p.e. of the approximation is $\dfrac{{}^+1}{{}^+64}$,
   (4) the r.e. of the approximation is $\dfrac{{}^+1}{{}^+64} \div {}^+2\frac{5}{32}$ or $\dfrac{{}^+1}{{}^+138}$.

We wish to call attention to the different information conveyed by

$$m(\overline{AB}) \doteq {}^+2, \qquad m(\overline{AB}) \doteq {}^+2.0, \qquad \text{and} \qquad m(\overline{AB}) \doteq {}^+2.00.$$

From the number of digits to the right of the decimal point we can tell that these three approximations are, respectively, correct to the nearest whole unit, $\frac{1}{10}$ of a unit, and $\frac{1}{100}$ of a unit.

The scheme unfortunately does not give unambiguous information in the following type of situation. If a friend tells you that the distance between Chicago and Los Angeles is approximately 2100 miles, you cannot tell definitely whether this means correct to the nearest mile, 10 miles, or 100 miles. Sometimes we attempt to clarify this information by saying something like "the distance is 2100 miles plus or minus 100 miles" or ". . . give or take a 100 miles"; this indicates that our approximation is correct to the nearest 100 miles. The situation is that in the approximation of $^+2100$ miles the two digits "0" do not give us any information about the precision of the approximation; they are referred to as *nonsignificant* digits; on the other hand, the "1" does tell us that the distance is between $^+2000$ miles and $^+2200$ miles; it is called a *significant* digit. Precise rules for determining which digits of a numeral are significant exist; we shall not pursue these at length, however, we take a brief look at a notational scheme called *scientific notation* which conveniently gives this information. Using scientific notation we write

$^+2100$ as $^+2.1 \times {}^+10^3$

to indicate that the approximation is correct to the nearest 100 miles (that is, neither "0" is significant); we write

$^+2100$ as $^+2.10 \times {}^+10^3$

to indicate that the approximation is correct to the nearest 10 miles (that is, the "0" following "1" in "$^+2100$" is significant).

The name, "scientific notation," indicates its rather special usage; therefore, we will not elaborate upon it in this text.

A remark about length   In the practice of approximating measures of models of line segments we use units of measure such as inch, foot, yard, mile, meter, centimeter, etc. If we approximate the inch-measure (that is, the measure with respect to an inch-unit) to be $^+17$, then we report the *length* of $\overline{AB}$ to be approximately $^+17$ inches. [A measure of a line segment (and an approximate measure of a model of a line segment) is a positive real number; to report the length we give both the measure (positive real number) and the unit used to determine the measure.] The approximate measure of the model of $\overline{AB}$ using a foot-unit would be $^+1\frac{5}{12}$; we would say its length is $^+1\frac{5}{12}$ feet. Since the measure of line segments (and of models of line segments) depends on the unit, every line segment (and every model of a line segment) has many measures. Do line segments (and also models of line segments) have more than one length? (Think carefully and review the preceding postulates!) Pursuing the illustration using the length of the model of $\overline{AB}$ will suggest an answer. Recall that a measure of $^+1$ with respect to a foot-unit equals $\dfrac{{}^+1}{{}^+12}$ times a measure of

$^{+}12$ with respect to an inch-unit; that is, we can "convert" an inch-measure of $^{+}17$ to a foot-measure by multiplying $\frac{+1}{+12} \times {}^{+}17$; therefore, an inch-measure of $^{+}17$ equals a foot-measure of $^{+}1\frac{5}{12}$. (Which postulate is an abstraction of the "conversion" illustrated above?) It may be necessary for the reader to review Postulates 4.3–4.5. This suggests that line segments (and models of line segments) have many measures but exactly one length.

Exercise set 4.4

1. The accompanying models are of three line segments which are to serve as units of measure. The names assigned to these are of no significance other than to help illustrate the concepts of measure and length. Draw a copy of each of the model unit

Yule unit

Mule unit

Tule unit

line segments on a sheet of paper and then approximate the measure of the accompanying models of line segments AB and CD.

a. The approximate yule measure of $\overline{AB}$ $(m_Y(\overline{AB}) \doteq ?)$ is _____. The length of $\overline{AB}$ is approximately _____ yules.

b. $m_M(\overline{AB}) \doteq$ _____; therefore, the length of $\overline{AB}$ is approximately _____.

c. $m_T(\overline{AB}) \doteq$ _____; therefore, the length of $\overline{AB}$ is approximately _____.

d. $m_M(\overline{CD}) \doteq$ _____ and the length of $\overline{CD}$ is _____.
$m_T(\overline{CD}) \doteq$ _____ and the length of $\overline{CD}$ is _____.

2. For each of the following approximate measures complete sentences (1)–(4) below to make true statements.

a. $m(\overline{AB}) \doteq {}^{+}4.$ b. $m(\overline{AB}) \doteq {}^{+}4.0.$

c. $m(\overline{AB}) \doteq {}^{+}4.04.$ d. $m(\overline{AB}) \doteq {}^{+}4.040.$

(1) _____ $\leq m(\overline{AB}) <$ _____ .
(2) The precision of the approximation is _____ .
(3) The g.p.e. of the approximation is _____ .
(4) The r.e. of the approximation is _____ .

3. Explain the difference in meaning between reporting $m(\overline{CD}) \doteq$ $^+5$ and $m(\overline{CD}) \doteq {}^+5.0$.

4. For each of the approximate measures complete sentences (1)–(4) of Exercise 2.

   a. $m(\overline{AB}) \doteq {}^+6.61$.   b. $m(\overline{AB}) \doteq {}^+4.732$.
   c. $m(\overline{AB}) \doteq {}^+2\frac{1}{2}$.   d. $m(\overline{AB}) \doteq {}^+2\frac{1}{4}$.
   e. $m(\overline{AB}) \doteq {}^+2\frac{3}{8}$.   f. $m(\overline{AB}) \doteq {}^+3\frac{5}{64}$.

5. The approximation of a measure is $^+245{,}000$.

   a. If the approximation is correct to the nearest 1000 units, then:
      (1) Write this in scientific notation. (*Note:* A numeral in scientific notation is of the form $a \times {}^+10^n$, where a is a positive integer and $^+1 < a < {}^+10$, and n is an integer.)
      (2) Determine the precision of the approximation.
      (3) Determine the g.p.e. of the approximation.
      (4) Determine the r.e. of the approximation.
   b. Do (1)–(4) above if the approximation is correct to the nearest 100 units.
   c. Do (1)–(4) above if the approximation is correct to the nearest 10 units.

6. Let the following be units of measure.

        •————• Fule   •——————————• Bule
        A B C D

   a. $m_F(\overline{CD}) \doteq$ _____ .
   b. For every line segment XY, $m_F(\overline{XY}) =$ _____ $\times m_B(\overline{XY})$.
   c. Draw a bule ruler with a length of four bules.
   d. On your bule ruler shade the interval of uncertainty for a line segment XY, where $m_B(\overline{XY}) \doteq 3$.
   e. _____ $\leq m_B(\overline{XY}) <$ _____ .
   f. Subdivide your bule ruler using the fule unit.
   g. _____ $\leq m_F(\overline{XY}) <$ _____ .

h. Suppose $m_B(\overline{RS}) \doteq 7$, then

    (1) _____ $\leqslant m_B(\overline{RS}) <$ _____ , and

    (2) _____ $\leqslant m_F(\overline{RS}) <$ _____ .

i. Suppose that $\overline{PQ}$ is a line segment such that $m_B(\overline{PQ}) \doteq 5$. Is it necessarily true that $m_F(\overline{PQ}) \doteq 15$? (*Note:* "$\doteq$" is intended to mean that the approximation is given to the nearest unit.)

4.5 MEASURE OF ANGLES, CONGRUENT ANGLES

Our ideas about the measure of line segments evolved from the idea of distance between two points. That is, the notion of the measure of a line segment arose from the consideration of geometric figures (two points called a "unit pair") which determine a line segment. The line segment determined by a unit pair is called a *unit line segment*. After choosing two points, and thus a line segment, as a unit for comparison we were able to determine the measure of every given line segment.

To develop the notion of angle measure we start by choosing a pair of geometric figures which determine an angle; namely, a pair of noncollinear rays with a common endpoint. To such a pair of rays, which we call a "unit pair" (or "unit angle"), we assign the real number $^+1$; $^+1$ is called the *measure* of the unit angle. Then, to determine the measure of any given angle we compare the given angle to the unit angle. To determine a given angle's measure we must determine the number of unit angles such that the union of the unit angles and their interiors equals the union of the given angle and its interior. If, for example, the number of unit angles for a given angle is four, then the measure of the given angle is $^+4$.

As in the case of the measure of a line segment, the measure of physical models of angles is necessarily approximate. We illustrate this in Example 4.5.

EXAMPLE 4.5 The accompanying model shows a unit angle.

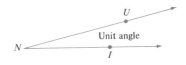

To determine the measure of $\angle ABC$ as suggested by the model

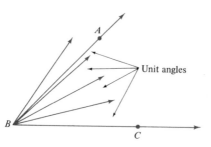

at the top of this page, we proceed as illustrated. We can see that the measure of ∠ABC is less than the measure of four unit angles and greater than the measure of three unit angles. Therefore, the measure of ∠ABC is between ⁺3 and ⁺4.

If, in Example 4.5, we would "subdivide" the unit angle into "subunit" angles, we could obtain a better approximation of the measure of ∠ABC. If we could subdivide the unit angle into an infinite set of "subunits," we could obtain the exact measure of ∠ABC.

What we are after in developing the ideas of angle measure is to define functions from the set of all angles to the set of real numbers. Our preceding discussion motivates and suggests how to define such functions. To do this we consider ray coordinate systems. To develop a *degree* ray coordinate system we proceed as follows:

We choose a pair of opposite rays; for the purpose at hand we choose a ray AB and its opposite ray AC; to ray AB we assign the real number 0, and to ray AC the real number ⁺180. Then, to every ray AX where X is in *one* of the half-planes determined by $\overleftrightarrow{AC}$, we assign a real number between 0 and ⁺180; these assignments define a one-to-one correspondence between this set of rays and the set of real numbers from 0 through ⁺180. Figure 4.2 suggests this one-to-one correspondence.

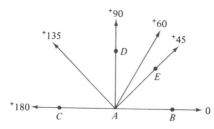

FIG. 4.2
Degree ray co-ordinate system.

The one-to-one correspondence suggested in Fig. 4.2 is called a *degree ray coordinate system*. Other types of ray coordinate systems are also used. The reader may be familiar with a radian coordinate system; with reference to Fig. 4.2, a correspondence between some rays and real numbers determined by a radian ray coordinate system is as follows:

$$\overrightarrow{AB} \leftrightarrow 0, \qquad \overrightarrow{AC} \leftrightarrow \pi, \qquad \overrightarrow{AD} \leftrightarrow \frac{\pi}{{}^{+}2}, \qquad \overrightarrow{AE} \leftrightarrow \frac{\pi}{{}^{+}4}.$$

A ray coordinate system based on a revolution would determine a correspondence as follows:

$$\overrightarrow{AB} \leftrightarrow 0, \qquad \overrightarrow{AC} \leftrightarrow \frac{{}^{+}1}{{}^{+}2}, \qquad \overrightarrow{AD} \leftrightarrow \frac{{}^{+}1}{{}^{+}4}, \qquad \overrightarrow{AE} \leftrightarrow \frac{{}^{+}1}{{}^{+}8}.$$

Hereafter we will use the *degree ray coordinate system* to assign to every angle a real number *between* 0 and $^{+}180$. This assignment defines a function whose domain is the set of all angles and range is $\{x \in \mathbb{R} | 0 < x < {}^{+}180\}$ and is called the *degree measure function*. For every $\angle ABC$, we denote the degree measure of $\angle ABC$ by $m_{\circ}(\angle ABC)$. That is, "$m_{\circ}(\angle ABC) = {}^{+}45$" is read "the degree measure of $\angle ABC$ is $^{+}45$." (Since our discussion will involve only the use of the degree coordinate system, we are justified in referring to *the* measure of an angle.) The method by which the real number is assigned to every angle is given in Definition 4.6.

Definition 4.6 For every angle ABC, the *degree measure* of $\angle ABC$, denoted by $m_{\circ}(\angle ABC)$, is

$$m_{\circ}(\angle ABC) = |x - y|, \qquad \text{where} \quad 0 < x < {}^{+}180, \quad 0 < y < {}^{+}180,$$

and x and y are the coordinates of $\overrightarrow{BA}$ and $\overrightarrow{BC}$ with respect to a degree ray coordinate system.

We illustrate Definition 4.6 in Example 4.6.

EXAMPLE 4.6 Using the model on p. 106 we can obtain degree measures for the following angles:

a. $\begin{aligned} m_{\circ}(\angle ABC) &= |{}^{+}60 - {}^{+}45| \\ &= |{}^{+}15| \\ &= {}^{+}15. \end{aligned}$

4 distance, measure, congruence of line segments and angles

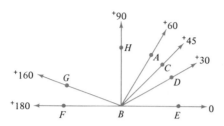

b. $m_\circ(\angle GBA) = |{}^+160 - {}^+60|$
$= |{}^+100|$
$= {}^+100.$

With respect to degree measure of angles there remains the question of how we know that there is always a degree coordinate system which is so conveniently located that the endpoint of the ray corresponding to zero is the same point as the vertex of the given angle. This is guaranteed in Postulate 4.6.

Postulate 4.6 (The protractor postulate) For every angle *ABC*, there exists a degree coordinate system such that the coordinate of $\overrightarrow{BA}$ is 0 and the coordinate of $\overrightarrow{BC}$ is a real number *x* and $0 < x < {}^+180$.

We now have the necessary geometric tools to define the relation of congruence on angles, which is given in Definition 4.7.

Definition 4.7 For all angles *ABC* and *DEF*, $\angle ABC$ is *congruent* to $\angle DEF$ if and only if $m_\circ(\angle ABC) = m_\circ(\angle DEF)$. If $\angle ABC$ is congruent to $\angle DEF$, we write $\angle ABC \cong \angle DEF$.

We state as a theorem some important properties of the relation $\cong$ on the set of all angles. The proof of the theorem is left as an exercise for the reader.

Theorem 4.3 The relation $\cong$ on angles has the following properties:

a. For every angle *ABC*, $\angle ABC \cong \angle ABC$.
b. For all angles *ABC* and *DEF*, if $\angle ABC \cong \angle DEF$, then $\angle DEF \cong \angle ABC$.

c. For all angles *ABC*, *DEF*, and *GHI*, if $\angle ABC \cong \angle DEF$ and $\angle DEF \cong \angle GHI$, then $\angle ABC \cong \angle GHI$.

That is, $\cong$ is an equivalence relation on the set of all angles.

4.6 CLASSIFICATION OF ANGLES, PERPENDICULARITY

The notions of angle measure and congruence provide a means for a classification of angles. The reader is undoubtedly familiar with many of the classifications.

An angle is a *right angle* if and only if its degree measure equals $^+90$.

An angle is an *acute angle* if and only if its degree measure is less than $^+90$.

An angle is an *obtuse angle* if and only if its degree measure is greater than $^+90$.

[*Note:* Keep in mind that for every angle *ABC*, $0 < m_\circ(\angle ABC) < {}^+180$.]

These classifications are illustrated in Example 4.7.

EXAMPLE 4.7 Parts (a), (b), and (c) of the accompanying model suggest right $[m_\circ(\angle ABC) = {}^+90]$, acute $[m_\circ(\angle DEF) < {}^+90]$, and obtuse $[m_\circ(\angle GHI) > {}^+90]$ angles, respectively.

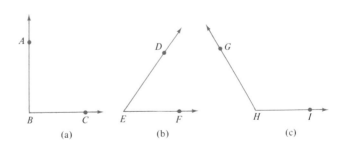

(a)   (b)   (c)

Two angles are *complementary* if and only if the sum of their degree measures equals $^+90$.

Two angles are *supplementary* if and only if the sum of their degree measures equals $^+180$.

The concept of right angle is useful in defining perpendicularity.

Definition 4.8   Two lines AB and CD are *perpendicular* if and only if $\overleftrightarrow{AB} \cup \overleftrightarrow{CD}$ is four right angles.   If $\overleftrightarrow{AB}$ is perpendicular to $\overleftrightarrow{CD}$, we write $\overleftrightarrow{AB} \perp \overleftrightarrow{CD}$.

We wish also to define the concept of perpendicularity for: line segment and line segment, line segment and line, line segment and ray, ray and ray, and so on.   To decrease the number of independent definitions let us agree to call line segments, half lines, rays, and lines *continuous subsets of lines:*

Two continuous subsets of two lines $\mathscr{L}_1$ and $\mathscr{L}_2$ are *perpendicular* if and only if

1.   their intersection is not empty, and
2. $\mathscr{L}_1 \perp \mathscr{L}_2$.

Thus two line segments AB and CD are perpendicular if and only if $\overline{AB} \cap \overline{CD} \neq \varnothing$ and $\overleftrightarrow{AB} \perp \overleftrightarrow{CD}$.   In case $\overline{AB}$ is perpendicular to $\overline{CD}$, we write $\overline{AB} \perp \overline{CD}$.   Some of these ideas are illustrated in Example 4.8.

EXAMPLE 4.8

a.   The accompanying models suggest continuous subsets of lines which are perpendicular.

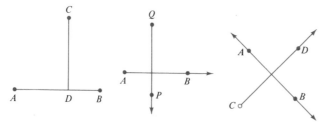

b.   The accompanying models suggest continuous subsets of lines which are not perpendicular.

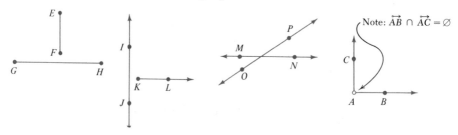

Definition 4.9 The *perpendicular bisector* of a line segment is the line which is perpendicular to the line segment at its midpoint.

Definition 4.10 A line $\mathcal{L}_1$ is *perpendicular* to a plane α at a point A if and only if $\mathcal{L}_1 \cap \alpha = \{A\}$ and for every line $\mathcal{L}$ in α which contains A, $\mathcal{L}_1 \perp \mathcal{L}$. If $\mathcal{L}_1$ is perpendicular to α, we write $\mathcal{L}_1 \perp \alpha$.

Definition 4.11 For all planes α and β, α is perpendicular to β if and only if $\alpha \cap \beta$ is a line and there exists a line $\mathcal{L}$ in α such that $\mathcal{L} \perp \beta$. If α is perpendicular to β, we write $\alpha \perp \beta$.

We leave to the reader the problem of defining the ideas of a line segment perpendicular to a plane, a ray perpendicular to a plane, and so on.

Exercise set 4.5

1. Prove Theorem 4.3.

2. In the accompanying diagram coordinates have been assigned to rays as shown.

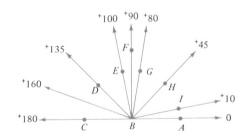

Indicate whether the statements below are true or false and complete the others to make true statements.

a. $m_\circ(\angle ABI) = |\underline{\hspace{1cm}} - \underline{\hspace{1cm}}| = \underline{\hspace{1cm}}$.
b. $m_\circ(\angle EBF) = |\underline{\hspace{1cm}} - \underline{\hspace{1cm}}| = \underline{\hspace{1cm}}$.
c. $\angle ABI = \angle EBF$.   d. $m_\circ(\angle CBD) = m_\circ(\angle FBH)$.
e. $\angle CBF = \angle FBA$.   f. $m_\circ(\angle FBD) = ?$.
g. $m_\circ(\angle FBH) = ?$.   h. $\angle FBD = \angle FBH$.
i. $\angle CBD$ and $\angle ABH$ are complementary.
j. $\angle ABG$ and $\angle GBF$ are complementary.
k. $m_\circ(\angle IBG) + m_\circ(\angle FBE) = ?$.
l. $\underline{\hspace{1cm}}$ is supplementary to $\angle ABF$.

4 distance, measure, congruence of line segments and angles

   m. $\angle HBD$ and $\angle ABF$ are complementary.
   n. _____ is supplementary to $\angle IBA$.
   o. $m_\circ(\angle IBF) = {}^+8 \cdot m_\circ(\angle ABI)$.
   p. $\angle EBG = {}^+2 \times \angle ABI$.

3. Review the definition of betweenness for points and note how it uses the measure of distance. Use the measure of angles to formulate a definition of betweenness for rays.

4. Let $\angle ABC \cong \angle DBE$, $\overset{\circ}{\overrightarrow{BD}} \subseteq \text{Int}(\angle ABC)$ and $\overset{\circ}{\overrightarrow{BC}} \subseteq \text{Int}(\angle DBE)$. Prove that $\angle ABD \cong \angle CBE$. (*Hint:* See the accompanying figure.)

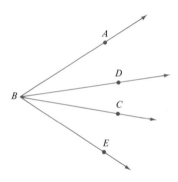

5. **a.** Suppose $\angle ABC \neq \angle DEF$. Does it necessarily follow that $\angle ABC \cong \angle DEF$? Explain your answer.
   b. Suppose $m_\circ(\angle ABC) > m_\circ(\angle DEF)$. Does it necessarily follow that $\angle ABC \cong \angle DEF$?. Explain your answer.
   c. Suppose $m_\circ(\angle ABC) > m_\circ(\angle DEF)$. Does it necessarily follow that $\angle ABC \neq \angle DEF$? Explain your answer.

6. Does the relation $\perp$ on line segments have:

   a. the reflexive property?   **b.** the symmetric property?
   c. the transitive property?

   Draw some models to illustrate.

7. Does the relation "is supplementary to" for angles have:

   a. the reflexive property?   **b.** the symmetric property?
   c. the transitive property?

   Draw some models to illustrate.

8. Repeat Exercise 7 for the relation "is complementary to" for angles.

4.7 TRANSVERSALS AND PARALLEL LINES

We have given a definition of parallel lines but we do not have a criterion other than to consider their intersection set to determine whether two lines are parallel. To give such a criterion is the major objective of this section. As is so often the case in geometry, to get to this criterion we need to introduce some additional vocabulary used in connection with lines and angles.

Two coplanar angles are *adjacent angles* if and only if they have the same vertex, one side in common, and no interior points in common.

Two coplanar angles are *vertical angles* if and only if their sides form two pairs of opposite rays.

These ideas are illustrated in Example 4.9.

EXAMPLE 4.9 Consider points, rays, angles, and lines suggested by the accompanying models. The following are true statements.

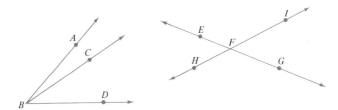

1. ∠DBC and ∠CBA are adjacent angles.
2. Although ∠DBA and ∠CBA have side $\overrightarrow{BA}$ in common, they are not adjacent angles. Why?
3. ∠GFI and ∠HFE are vertical angles.
4. ∠IFE and ∠HFG are vertical angles.
5. ∠EFI and ∠IFG are adjacent angles.

Theorem 4.4 If ∠ABC and ∠DBE are vertical angles, then ∠ABC ≅ ∠DBE.

The proof depends on Exercise 4 of Exercise Set 4.6 and is left as an exercise for the reader in Exercise 6 of Exercise Set 4.6.

Frequently when we are concerned with showing two lines parallel, there is a third line which can be used.

A line $\mathscr{L}_1$ is called a *transversal* of lines $\mathscr{L}_2$ and $\mathscr{L}_3$ if and only if $\mathscr{L}_1$, $\mathscr{L}_2$, and $\mathscr{L}_3$ are coplanar and $\mathscr{L}_1$ intersects $\mathscr{L}_2 \cup \mathscr{L}_3$ in exactly two points.

In Fig. 4.3 $\mathscr{L}_1$ is a transversal of $\mathscr{L}_2$ and $\mathscr{L}_3$ but $\mathscr{L}_4$ is not a transversal of $\mathscr{L}_5$ and $\mathscr{L}_6$.

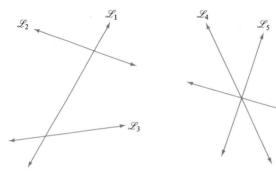

FIG. 4.3
Models to illustrate the concept of a transversal.

A transversal $\mathscr{L}_1$ forms with two lines $\mathscr{L}_2$ and $\mathscr{L}_3$ a total of eight angles. We classify the angles by pairs as follows.

Two coplanar angles are *alternate interior* angles if and only if their intersection set is a line segment and the intersection set of their interiors is the empty set.

Two coplanar angles are *consecutive interior* angles if and only if their intersection set is a line segment and the intersection set of their interiors is not the empty set.

Two coplanar angles are *corresponding* angles if and only if their intersection set is a ray and the intersection set of their interiors is not the empty set.

These ideas are illustrated in Example 4.10.

EXAMPLE 4.10 Consider the accompanying model of coplanar lines and a transversal.

$\angle ABC$ and $\angle ECB$ are alternate interior angles.
$\angle DCB$ and $\angle ABC$ are consecutive interior angles.
$\angle DCG$ and $\angle ABC$ are corresponding angles.

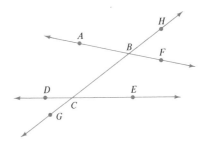

Which angle is the corresponding angle of ∠GCE? Which angle is the alternate interior angle of ∠DCB? Which angle is the consecutive interior angle of ∠ECB?

In a more complete development, the criterion for proving that two lines are parallel can be established from certain properties of triangles. Space limitations necessitate the omission of certain topics; therefore, we accept this for our development as an assumption which is Postulate 4.7.

Postulate 4.7 Two coplanar lines are parallel if and only if alternate interior angles formed by a transversal are congruent.

Exercise set 4.6

1. Let points, lines, and angles be related as suggested by the accompanying model.

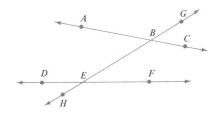

   a. Name all pairs of consecutive interior angles.
   b. Name all pairs of alternate interior angles.
   c. Name all pairs of corresponding angles.

2. Refer to the model on the next page. Name a pair of alternate interior angles.

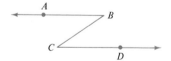

3. Refer to the accompanying model.

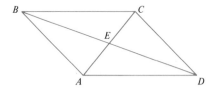

    a. Name two pairs of alternate interior angles.
    b. Name four pairs of consecutive interior angles.
    c. Name two pairs of vertical angles.

4. Prove: If the union of two sides of a pair of adjacent angles is a line, then the angles are supplementary. (A pair of adjacent angles such that the union of two of their sides is a straight line is called a *linear pair*.)

5. Prove: Two angles of a linear pair are congruent if and only if they are right angles.

6. Prove: Vertical angles are congruent.

7. Prove: If alternate interior angles formed by a transversal are congruent, then consecutive interior angles are supplementary.

8. Prove: If consecutive interior angles formed by a transversal are supplementary, then the alternate interior angles are congruent.

9. Prove: If alternate interior angles formed by a transversal are congruent, then corresponding angles are congruent.

10. Prove: If corresponding angles formed by a transversal are congruent, then alternate interior angles are congruent.

11. Prove: If two coplanar lines are perpendicular to the same line, they are parallel.

12. Prove: The perpendicular bisector of a line segment AB is the set of all points X such that $AX = XB$.

13. In the accompanying model:

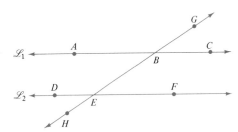

a. If $\mathcal{L}_1 \parallel \mathcal{L}_2$, then the alternate interior angles, $\angle ABE$ and $\angle BEF$, are _____ .

b. If $\angle FEB \cong \angle CBG$, then the alternate interior angles, $\angle FEB$ and $\angle ABE$, are _____ ; and, therefore $\mathcal{L}_1$ and $\mathcal{L}_2$ are _____ .

c. If $\angle FEB$ and $\angle EBC$ are supplementary, then the alternate interior angles, $\angle BED$ and $\angle EBC$, are _____ ; and, therefore $\mathcal{L}_1$ and $\mathcal{L}_2$ are _____ .

d. If $\mathcal{L}_1 \parallel \mathcal{L}_2$, then the alternate interior angles, $\angle ABE$ and $\angle BEF$, are _____ .

14. Complete and prove the following: Two coplanar lines are parallel if and only if corresponding angles formed by a transversal are _____ .

15. Complete and prove the following: Two coplanar lines are parallel if and only if consecutive angles formed by a transversal are _____ .

5

curves

5.1 DEFINITION AND CLASSIFICATION OF CURVES

The intuitive concept of a curve is very simple; therefore, it is
interesting that an accurate definition of a curve is rather difficult.
If you were placed "on the spot" to answer the question of what
is a curve, you would probably reply that a curve is something
which looks like the following model. Would you call the

following a model of a curve? By now you have gained

enough of an appreciation for the demands in mathematics to
realize that such a description of a curve comes far short of being
a definition. We will give a formal definition of a curve; however, it involves some concepts which we have not yet considered.

Does it surprise you that a curve can be defined using the concept of a function? More specifically a curve is the set of image

points of a function whose domain is a unit line segment OI, and range is a subset of space. However (and this is where our difficulty comes), not every function from $\overline{OI}$ to S is a curve. We need to restrict the function. Let us look at diagrams of some functions from $\overline{OI}$ to a subset of space and think about whether we would want to call the set of image points a curve. In Figs. 5.1(a) and

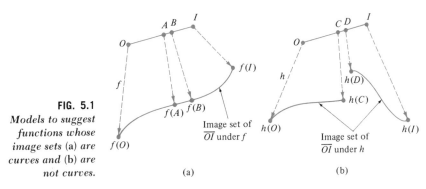

FIG. 5.1
Models to suggest functions whose image sets (a) are curves and (b) are not curves.

(a)   (b)

(b), let us concentrate on points A and B, and C and D, and their images $f(A)$ and $f(B)$, and $h(C)$ and $h(D)$, respectively. In both cases there is probably agreement that the points A and B, and C and D, respectively, are "close together." Moreover, we can probably agree that the points $f(A)$ and $f(B)$ *are* "close together" but that $h(C)$ and $h(D)$ *are not* "close together." This intuitively characterizes the idea of a *continuous* function from $\overline{OI}$ to S. That is, a function f from $\overline{OI}$ to S is *continuous* if and only if points in $\overline{OI}$ which are "close together" are mapped to points in S which are "close together." It still remains rather vague by what is meant by "close together." Unfortunately, space limitations of this text prohibit pursuing this notion to the point of defining it precisely. The careful study of the meaning and properties of continuous functions constitutes a substantial portion of an area of mathematics called the "calculus."

Definition 5.1 A *curve* in S is the set of image points of a continuous function f from the unit line segment to S. The points $f(O)$ and $f(I)$ are called *initial* and *terminal* points of the curve, respectively.

The model in Fig. 5.2 (next page) illustrates Definition 5.1.

5 curves

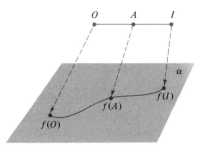

FIG. 5.2
Model of a curve.

Definition 5.2 A curve $\mathscr{C}$ is a *plane curve* if and only if there exists a plane α such that $\mathscr{C} \subseteq \alpha$.

The model in Fig. 5.3 illustrates Definition 5.2.

FIG. 5.3
Model of a plane curve.

Hereafter in this text, the word "curve" will mean *plane curve;* that is, our study of curves will be limited to those curves which are subsets of a plane.

In Fig. 5.4 are models of sets of image points of functions from

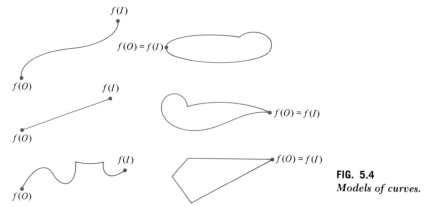

FIG. 5.4
Models of curves.

$\overline{OI}$ to a plane α; that is, models of curves. What we hope the reader has observed from the models is that for some curves, but not for all, $f(O) = f(I)$. That is, for some curves the initial and terminal points are the same point. If $f(O) \neq f(I)$ and for every point $A \in \overline{OI}$, $A \neq O$ and $A \neq I$, $f(O) \neq f(A)$ and $f(I) \neq f(A)$, then the points $f(O)$ and $f(I)$ are called *endpoints*. We classify curves according to whether $f(O) = f(I)$ and refer to those curves in which $f(O) = f(I)$ as being *closed curves*.

Definition 5.3 A *closed curve* is the set of image points of a continuous function f from the unit line segment OI to a plane α, such that $f(O) = f(I)$.

That is, for a closed curve, the initial point and terminal point coincides. The reader should refer to the models of curves in Fig. 5.4 and identify those in which: (a) $f(O)$ and $f(I)$ are endpoints, and (b) $f(O) = f(I)$; that is, identify the models of closed curves.

The models in Fig. 5.5 illustrate Definition 5.3.

FIG. 5.5
Models of closed curves.

The models in Fig. 5.6 suggest another way to classify curves. Models (a), (b), and (c) suggest that there exist points $A, B \in \overline{OI}$, not both endpoints, such that $f(A) = f(B)$. Whereas, models (d) and (e) suggest that for all points $A, B \in \overline{OI}$, not both endpoints, $f(A) \neq f(B)$. We refer to models (d) and (e) as models of *simple* curves; since (e) is also a model of a closed curve, we call the curve a *simple closed* curve.

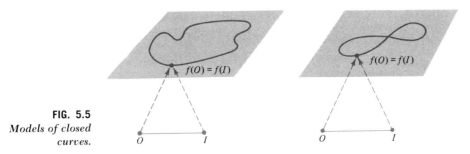

FIG. 5.6
Models to suggest a classification of curves.

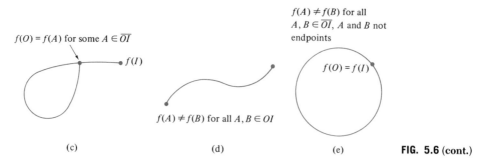

(c) (d) (e) **FIG. 5.6** (cont.)

Definition 5.4 A *simple* curve is the set of image points of a continuous function f from $\overline{OI}$ to a plane α such that for all points $A,B \in \overline{OI}$, not both endpoints of $\overline{OI}$, $f(A) \neq f(B)$.

Definition 5.5 A *simple closed* curve is the set of image points of a continuous function f from $\overline{OI}$ to a plane α, such that $f(O) = f(I)$ and for all points $A,B \in \overline{OI}$, not both endpoints of $\overline{OI}$, $f(A) \neq f(B)$.

5.2 SOME SPECIAL SIMPLE CLOSED CURVES, INTERIOR AND EXTERIOR

Some closed curves, and physical models of these curves, occur so frequently that we give them special names. It surely comes as no surprise to the reader that a circle is a simple closed curve.

Definition 5.6 A *circle* is a simple closed curve in a plane α for which there exist a point P in plane α and a real number r, $r > 0$, such that for every point Q on the curve, the distance between P and Q equals r. The point P is called the *center* of the circle and the positive real number r is called the *radius* of the circle.[1]

We wish to emphasize that a circle is the *set of all points* in a plane such that the measure of the distance from every point on the curve to a point P is the same. Symbolically, a circle $\mathscr{C}$ is the following set:

$$\mathscr{C} = \{X \in \alpha |\text{ there exists a point } P \in \alpha \text{ and there exists}$$
$$r \in \mathbb{R}, r > 0, \text{ such that } XP = r\}.$$

---

[1] Some authors define a radius to be a line segment joining the center and a point on the circle; other authors use "radius" to denote both the distance and the line segment.

Postulate 5.1   For every point $P \in S$ and for every real number r, $r > 0$, there is a circle with center at P and radius r.

We asked the reader to think about whether a line segment is a curve; the answer is yes. Also, the union of two curves which have a common endpoint is a curve. Many closed curves are sets of points which are unions of line segments.

Definition 5.7   A *polygon* is a simple closed curve which is the union of line segments. Each of the line segments is a *side* of the polygon and each of the endpoints of the line segments is a *vertex* of the polygon.

In Fig. 5.7, models of simple closed curves $\mathscr{C}_1$ and $\mathscr{C}_2$ are shown. The models suggest that points P' and Q' and P and Q are related to the simple closed curves $\mathscr{C}_1$ and $\mathscr{C}_2$, respectively, in the same

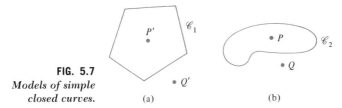

FIG. 5.7
*Models of simple
closed curves.*   (a)   (b)

way that points in opposite half-planes are related to the line which is the separating line; and also in the way two points, one in the interior and one in the exterior of an angle are related to the angle. The models suggest Postulate 5.2.

Postulate 5.2   A simple closed curve $\mathscr{C}$ separates the set of points in a plane α which are not on $\mathscr{C}$, into two disjoint subsets such that if P and Q are not on $\mathscr{C}$ and are not in the same subset, then $\overline{PQ} \cap \mathscr{C} \neq \varnothing$.

One of the two subsets determined by a simple closed curve $\mathscr{C}$ in a plane α is the interior and the other the exterior. Our models suggest which is the interior and which is the exterior. We will give a general definition of the interior of simple closed curves. After we give an illustration of the general definition we will give a criterion which characterizes the interior and the exterior of *most* curves; "most" in this context means *all* for which one can draw a model.   For both the general definition of the interior and the exterior of a simple closed curve and their characterizations for *most* curves we need Definition 5.8.

Definition 5.8   For every circle $\mathscr{C}$ in a plane α with center P and radius r, the *interior* of $\mathscr{C}$ is the set of all points $X \in \alpha$ such that $XP < r$.   We denote the interior of $\mathscr{C}$ by $\text{Int}(\mathscr{C})$.

The model in Fig. 5.8 suggests the interior and exterior of a circle.

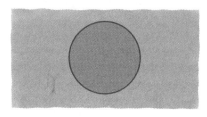

FIG. 5.8
Model to suggest interior and exterior of a circle.

Definition 5.9   For every set $\mathscr{A}$ in a plane α, $\mathscr{A}$ is *bounded* if and only if there exists a circle $\mathscr{C}$ such that $\mathscr{A} \subseteq \text{Int}(\mathscr{C})$.

Some of the following questions refer to the models in Fig. 5.9 and should help to clarify Definition 5.9.   Is a line segment a

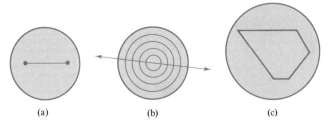

(a)   (b)   (c)

FIG. 5.9
Models to suggest bounded and unbounded sets.

bounded set?   [See model (a) of Fig. 5.9.]   Is a line a bounded set?   [See model (b) of Fig. 5.9.]   The interior of an angle?   A polygon?   A circle?

Postulate 5.3   Let $\mathscr{C}$ be a simple closed curve in a plane α.   One of the disjoint subsets of the points not on $\mathscr{C}$, guaranteed by Postulate 5.2, is bounded, the other is not.

The models in Fig. 5.10 illustrate Postulate 5.3.

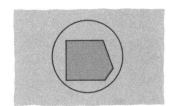
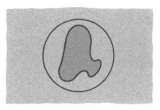

FIG. 5.10
Models to illus-
trate Postulate 5.3.

Definition 5.10   For every simple closed curve $\mathscr{C}$ in a plane α, the *interior* of $\mathscr{C}$, denoted by Int($\mathscr{C}$), is the bounded set guaranteed by Postulate 5.3.

Definition 5.11   For every simple closed curve $\mathscr{C}$ in a plane α, the *exterior* of $\mathscr{C}$, denoted by Ext($\mathscr{C}$), is the unbounded set guaranteed by Postulate 5.3.

Definitions 5.10 and 5.11 are illustrated in Example 5.1.

EXAMPLE 5.1

a. Consider the accompanying models of simple closed curves. The regions shaded are the interiors, because as illustrated, there exist circles whose interior contains the shaded region.

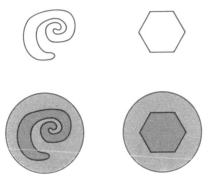

b. Consider the model (next page) of a simple closed curve. We employ the "shading technique" to determine whether P is in the interior or exterior of the curve.   Since the shaded region clearly is part of the subset of a plane which cannot be contained in the interior of a circle, P is in the exterior of the given simple closed curve.

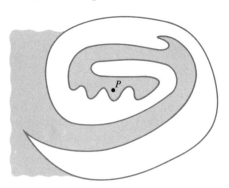

The characterization of the interior and exterior that we wish to give for *most* curves reduces to a simple process of counting the number of times a particular ray "crosses" the curve. We first explain precisely what we mean when we say that a ray crosses a curve.

A ray AB *crosses* a simple curve $\mathscr{C}$ at a point D if and only if:

a. $D \in \overrightarrow{AB} \cap \mathscr{C}$,

b. $D \neq A$, and

c. the interior of every circle with center at D contains points of $\mathscr{C}$ in *both* of the opposite half-planes of $\overleftrightarrow{AB}$.

The models in Example 5.2 suggest how to apply the above statement to models of rays and curves.

EXAMPLE 5.2

a. In the accompanying model $\overrightarrow{AB}$ crosses curve $\mathscr{C}$ at D because $D \in \overrightarrow{AB} \cap \mathscr{C}$, $D \neq A$, and the interior of every circle with center at D contains points of the curve in both of the opposite half-planes of $\overleftrightarrow{AB}$.

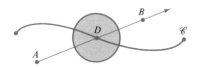

b. In the accompanying model $\overrightarrow{AB}$ does not cross curve $\mathscr{C}$ because $\{A\}$ is the intersection set $\overrightarrow{AB} \cap \mathscr{C}$.

c. In the accompanying model $\overrightarrow{AB}$ does not cross curve $\mathscr{C}$ at D because there exists a circle such that all of the points on the curve which are interior to the circle are "on the same side of $\overleftrightarrow{AB}$."

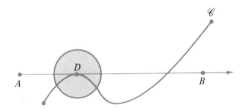

The model in Example 5.3 suggests another method, for most curves, to determine whether a point is in the interior or exterior.

EXAMPLE 5.3 As illustrated by the accompanying model, there is a ray with endpoint P which crosses the curve in an *odd* number of points; from this it is possible to conclude that P is in the interior. There is a ray with endpoint Q which crosses the curve in

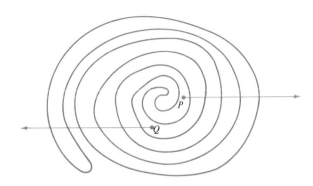

an *even* number of points; from this it is possible to conclude that Q is in the exterior.

Example 5.3 above suggests a characterization of the interior and exterior of *most* simple closed curves which we state as follows:

For *most* simple closed curves $\mathscr{C}$, and for every point X, $X \notin \mathscr{C}$, $X \in \text{Int}(\mathscr{C})$ if and only if there exists a ray with endpoint X which crosses the curve in an odd number of points.

For *most* simple closed curves $\mathscr{C}$, and for every point X, $X \notin \mathscr{C}$, $X \in \text{Ext}(\mathscr{C})$ if and only if there exists a ray with endpoint X which crosses the curve in an even number of points.

The concept of the interior of a curve provides another way to classify simple closed curves.

Definition 5.12 A simple closed curve $\mathscr{C}$ is a *convex* curve if and only if for all points $X, Y \in \mathscr{C}$, $\overline{XY} \subseteq \mathscr{C} \cup \text{Int}(\mathscr{C})$.

We can see from Definition 5.12 that a simple closed curve is *convex* if and only if its interior is a convex set.

The concept of a convex curve is illustrated by the models in Example 5.4.

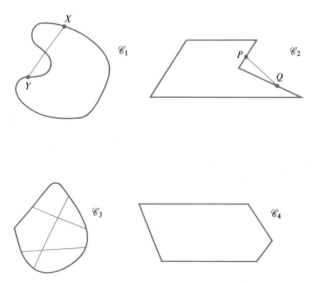

EXAMPLE 5.4 Consider the curves suggested by the preceding models. Since there exist points X and Y on $\mathscr{C}_1$, such $\overline{XY} \subseteq \mathscr{C}_1 \cup \text{Int}(\mathscr{C}_1)$, $\mathscr{C}_1$ is not a convex curve. Does it appear that $\mathscr{C}_3$ and $\mathscr{C}_4$ are convex curves? Are $\mathscr{C}_3$ and $\mathscr{C}_4$ *convex sets?* (Be careful!) Is $\mathscr{C}_2$ a convex curve? Explain. Is it a convex set?

Exercise set 5.1

1. For the accompanying models of curves indicate which of the following apply:

   simple,   polygon,
   closed,   convex curve,
   simple closed,   convex set.

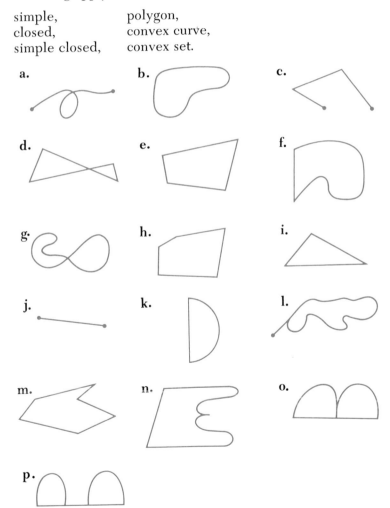

a.

b.

c.

d.

e.

f.

g.

h.

i.

j.

k.

l.

m.

n.

o.

p.

2. Let $\mathcal{C}$ be a circle with radius r and center P. Indicate whether the following are true or false.

a. $P \in \mathcal{C}$. b. $r \in \mathcal{C}$.
c. $P \in \text{Int}(\mathcal{C})$. d. $P \in \text{Ext}(\mathcal{C})$.
e. r is a positive real number. f. r is a set of points.

3. Indicate whether the accompanying models suggest that the ray AB crosses the curve $\mathcal{C}$ at the point D. Explain your answer.

a.

b.

c.

d.

e.

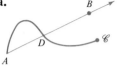

f.

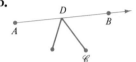

4. Use both the shading technique and the counting technique to determine whether the points labeled P and Q in the accompanying models belong to the interior or the exterior of the model curves.

a.

b.

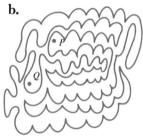

c.   **d.**

e.   **f.**

5. In the accompanying figure, part of the model for the simple closed curve $\mathscr{C}$ has been cut off. If point P is an element of the exterior of the curve $\mathscr{C}$, is point T an element of the interior of the curve $\mathscr{C}$? Explain your answer.

5.3 MEASURE OF CURVES

We have indicated that a line segment is a curve and have defined its measure as the distance between its endpoints. Since a line segment is a curve, we are really forced to agree that the union of two or more line segments, each of which has at least one point in common with another, is a curve. That is, we must agree that

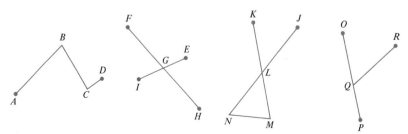

FIG. 5.11
Models of curves.

the models in Fig. 5.11 are models of curves. The question arises as to what to call such curves; for purposes of this text we answer the question in Definition 5.13.

Definition 5.13 A curve which is the union of line segments such that every line segment has exactly one point in common with at least one other line segment is a *union-of-line-segments curve*.

Now the question of what is the measure of a union-of-line-segments curve is easily answered and is given in Definition 5.14.

Definition 5.14 The *measure* of a union-of-line-segments curve is the sum of the measures of the line segments in the union.

Two illustrations of Definition 5.14 are given in Example 5.5.

EXAMPLE 5.5 Consider the accompanying models of union-of-line-segments curves $\mathscr{C}_1$ and $\mathscr{C}_2$.

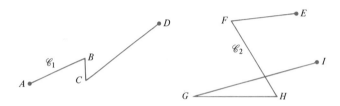

The measure of $\mathscr{C}_1$, $m(\mathscr{C}_1) = m(\overline{AB}) + m(\overline{BC}) + m(\overline{CD})$.
The measure of $\mathscr{C}_2$, $m(\mathscr{C}_2) = m(\overline{EF}) + m(\overline{FH}) + m(\overline{HG}) + m(\overline{GI})$.

There are many curves which are not union-of-line-segments curves and we cannot avoid meeting the question of what is meant by the measure of such a curve. What we need is some method to assign to every curve a unique real number which is called its measure. Techniques of the branch of mathematics called calculus are necessary to answer this question precisely.

We will give, with the aid of models, an intuitive notion of this technique. Consider the model of curve $\mathscr{C}$ shown in Fig. 5.12.

FIG. 5.12

Models to suggest the measure of a curve.

The measure of line segment AB $[m(AB)]$, although probably a poor one, is an approximation of the measure of $\mathscr{C}$; $m(\overline{AE}) + m(\overline{EF}) + m(\overline{FB})$ is probably a better approximation; and $m(\overline{AI}) + m(\overline{IJ}) + m(\overline{JK}) + m(\overline{KL}) + m(\overline{LM}) + m(\overline{MB})$ is better still. If we were to continue in this manner and let the number of line segments become greater and greater, we would approach the exact measure of the curve.

We next direct our attention to the measure of simple closed curves. A polygon is a union-of-line-segments curve which is simple and closed. The *measure of a polygon* is the sum of the measures of its sides. The word *perimeter* is related to polygons as length is related to line segments and curves; that is, the *perimeter* of a polygon is the measure of the polygon (a positive real number) together with the unit used to determine the measure.

Concepts of calculus make it possible to determine the measure of many simple closed curves. In particular, using concepts of calculus, it is possible to prove as a theorem that the measure of a circle is equal to $^{+}2\pi r$, where r is the radius of the circle. The *circumference* of a circle is a measure of the circle together with the unit used to determine the measure.

Exercise set 5.2

1. Determine the measure of the accompanying union-of-line-segments curves.

a.

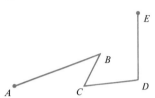

b.

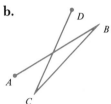

$$m(\overline{AB}) = {}^+3.$$
$$m(\overline{BC}) = \frac{{}^+3}{{}^+2}.$$
$$m(\overline{CD}) = {}^+2.$$
$$m(\overline{DE}) = \frac{{}^+5}{{}^+4}.$$

$$m(\overline{AB}) = m(\overline{BC})$$
$$= m(\overline{CD}) = {}^+5.$$

2. Determine the measure of the circle with radius r when:

   a. $r = {}^+1.$ b. $r = {}^+2.$

   c. $r = {}^+4.$ d. $r = {}^+8.$

   e. If r_1 and r_2 are the radii of circles $\mathscr{C}_1$ and $\mathscr{C}_2$, respectively,
   and $\dfrac{r_1}{r_2} = \dfrac{{}^+1}{{}^+2}$, then $\dfrac{m(\mathscr{C}_1)}{m(\mathscr{C}_2)} = ?.$

3. Approximate the measure of the circles described in Exercise
   2 parts a–d using ${}^+3.14$ as an approximation of π.

4. Consider the polygons suggested by the accompanying models.

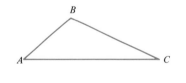

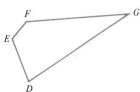

   a. Determine the measure of $\triangle ABC$ when $m(\overline{AB})$, $m(\overline{BC})$,
   and $m(\overline{CA})$ are, respectively:

   (1) ${}^+2, {}^+3, {}^+4$ (2) ${}^+3.9, {}^+6.4, {}^+9.3.$

   b. Determine the measure of quadrilateral $DEFG$ when
   $m(\overline{DE})$, $m(\overline{EF})$, $m(\overline{FG})$, and $m(\overline{GD})$ are, respectively:

   (1) ${}^+2, {}^+1, {}^+3, {}^+4$ (2) ${}^+1.2, {}^+1.7, {}^+6.9, {}^+7.3.$

5.4   THE PRACTICE OF APPROXIMATING THE MEASURE OF UNION–OF–LINE–SEGMENTS CURVES

In order to determine the measure of union-of-line-segments curves (and hence of polygons), we must add the measures of the line segments which are subsets of the curve; similarly, when we approximate the measure of models of such curves, addition of approximate measures is involved.   A natural question is whether the approximate measure of the model of the union-of-line-segments curve determined by this addition is as "good" as the approximation of the measures of the subsets.   Or more precisely, does the addition of approximate measures result in an approximation with the same greatest possible error (g.p.e.) and relative error (r.e.)?

Consider the model of a union-of-line-segments curve $\mathscr{C}$ shown in Fig. 5.13.   Suppose we determine that $m(\overline{AB}) \doteq {}^+3$ and $m(\overline{BC}) \doteq {}^+5$.   Then,

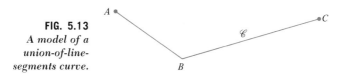

FIG. 5.13
A model of a union-of-line-segments curve.

$${}^+2.5 \leqslant m(\overline{AB}) < {}^+3.5 \quad \left(\text{g.p.e.} = {}^+0.5, \quad \text{r.e.} = \frac{{}^+1}{{}^+6}\right)$$

and

$${}^+4.5 \leqslant m(\overline{BC}) < {}^+5.5 \quad \left(\text{g.p.e.} = {}^+0.5, \quad \text{r.e.} = \frac{{}^+1}{{}^+10}\right).$$

Using the fact that $m(\mathscr{C}) = m(\overline{AB}) + m(\overline{BC})$ and properties of inequalities we have that

$${}^+7.0 \leqslant m(\mathscr{C}) < {}^+9.0 \quad \left(\text{g.p.e.} = {}^+1, \quad \text{r.e.} = \frac{{}^+1}{{}^+8}\right).$$

$\left(\textit{Note:} \quad {}^+1 = {}^+0.5 + {}^+0.5; \text{ that is, in this case the g.p.e. of the sum}\right.$

is the sum of the g.p.e.'s; also note that $\left.\dfrac{{}^+1}{{}^+10} < \dfrac{{}^+1}{{}^+8} < \dfrac{{}^+1}{{}^+6}.\right)$

Suppose that we determined that $m(\overline{AB}) \doteq {}^+3.5$ and that $m(\overline{BC}) \doteq {}^+5.5$.   Then,

$${}^+3.45 \leqslant m(\overline{AB}) < {}^+3.55 \quad \left(\text{g.p.e.} = {}^+0.05, \quad \text{r.e.} = \frac{{}^+1}{{}^+70}\right)$$

and

$$^+5.45 \leqslant m(\overline{BC}) < {}^+5.55 \qquad \left(\text{g.p.e.} = {}^+0.05, \quad \text{r.e.} = \frac{^+1}{^+110}\right);$$

so that

$$^+8.90 \leqslant m(\mathscr{C}) < {}^+9.10 \qquad \left(\text{g.p.e.} = {}^+0.10, \quad \text{r.e.} = \frac{^+1}{^+90}\right).$$

(*Note:* $^+0.10 = {}^+0.05 + {}^+0.05$; that is, again the g.p.e. of the sum of the two approximations is the sum of the g.p.e.'s of the two approximations. Moreover, $\dfrac{^+1}{^+110} < \dfrac{^+1}{^+90} < \dfrac{^+1}{^+70}$; that is, in this case, as in the previous example, the r.e. of the sum of two approximations is between the r.e.'s of the two approximations.)

We can prove that the above statements about g.p.e. and r.e. are true in general: Suppose that $\mathscr{C} = \overline{EF} \cup \overline{CD}$ is a union-of-line-segments curve, that $m(\overline{EF}) \doteq x$ with g.p.e. $= a$ $\left(\text{and hence r.e.} = \dfrac{a}{x}\right)$, and that $m(\overline{CD}) \doteq y$ with g.p.e. $= b$ $\left(\text{and hence r.e.} = \dfrac{b}{y}\right)$. Then,

$$x - a \leqslant m(\overline{EF}) < x + a,$$
$$y - b \leqslant m(\overline{CD}) < y + b.$$

From the fact that $m(\mathscr{C}) = m(\overline{EF}) + m(\overline{CD})$, and from the properties of inequalities it follows that

$$(x + y) - (a + b) \leqslant m(\mathscr{C}) < (x + y) + (a + b);$$

therefore, this approximation of $m(\mathscr{C})$ has a g.p.e. of $a + b$; that is, the g.p.e. of the approximation obtained by adding the two approximations is the sum of the g.p.e.'s of the addends.

Since the r.e. of the approximation $m(\mathscr{C}) \doteq x + y$ is $\dfrac{a + b}{x + y}$, to prove our conjecture that the r.e. of an approximation determined by adding two approximations is between the r.e.'s of the two approximations we must show that $\dfrac{a + b}{x + y}$ is between $\dfrac{a}{x}$ and $\dfrac{b}{y}$. Let us assume that $\dfrac{a}{x} \leqslant \dfrac{b}{y}$; then, we must show that

$$\frac{a}{x} \leqslant \frac{a + b}{x + y} \leqslant \frac{b}{y}.$$

Since

$$\frac{a}{x} \leq \frac{b}{y},$$

we have

$ay \leq bx$ (since a, b, x, and y are all positive).

$\therefore ax + ay \leq ax + bx.$

$\therefore a(x + y) \leq x(a + b).$

$$\therefore \frac{a}{x} \leq \frac{a + b}{x + y}.$$

We leave to the reader the problem of showing that

$$\frac{a + b}{x + y} \leq \frac{b}{y}.$$

An illustration of the application of these ideas to determining the measure of a model of a polygon is given in Example 5.6.

EXAMPLE 5.6 Consider the accompanying model of polygon $ABCD$ (curve $\mathscr{C}$). Suppose that the approximate measures of $\overline{AB}$, $\overline{BC}$, $\overline{CD}$, and $\overline{AD}$ with respect to an inch unit are

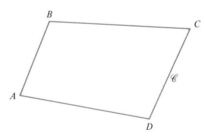

$m(\overline{AB}) \doteq {}^{+}3.50,$
$m(\overline{BC}) \doteq {}^{+}6.25,$
$m(\overline{CD}) \doteq {}^{+}4.35,$
$m(\overline{AD}) \doteq {}^{+}5.75.$

Since $m(\mathscr{C}) = m(\overline{AB}) + m(\overline{BC}) + m(\overline{CD}) + m(\overline{AD})$, it follows that $m(\mathscr{C}) \doteq {}^{+}3.50 + {}^{+}6.25 + {}^{+}4.35 + {}^{+}5.75$; that is,

$m(\mathscr{C}) \doteq {}^{+}19.85.$

That is, the measure of polygon $ABCD$ is approximately ${}^{+}19.85$ and the *perimeter* of polygon $ABCD$ is approximately ${}^{+}19.85$ *inches*. Since the g.p.e. of each of the approximations is ${}^{+}0.05$, the g.p.e.

of the approximation of the measure of polygon *ABCD* is ⁺0.20. Make a guess as to the r.e. of the approximation of the measure of polygon *ABCD*, then compute it to test your conjecture, and finally, state a generalization which is suggested by this observation.

Exercise set 5.3

1. Consider the accompanying line segments as units of measure with names as suggested.

    Blo _____
    Glo _____

    A. If $\overline{AB}$ is any line segment, let $m_B(\overline{AB})$ and $m_G(\overline{AB})$ denote the Blo and Glo measures of $\overline{AB}$, respectively.

    a. Let $\overline{AB}$ be a line segment such that $m_G(\overline{AB}) = {}^+1$, determine $m_B(\overline{AB})$.

    b. Let $\overline{CD}$ be a line segment such that $m_B(\overline{CD}) = {}^+1$, determine $m_G(\overline{CD})$.

    c. For every line segment *EF*, $m_B(\overline{EF}) = $ _____ $\times$ $m_G(\overline{EF})$, and $m_G(\overline{EF}) = $ _____ $\times m_B(\overline{EF})$.

    d. Construct a Blo ruler and a Glo ruler on edges of a sheet of paper.

    B. Consider the accompanying models of union-of-line-segments curves.

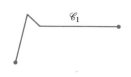

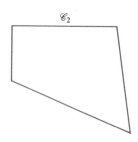

$\mathscr{C}_2$

$\mathscr{C}_1$

    a. (1) $m_G(\mathscr{C}_1) \doteq $ _____; therefore,
        (2) the length of $\mathscr{C}_1$ is approximately _____ Glos.
        (3) $m_B(\mathscr{C}_1) \doteq $ _____; therefore,
        (4) the length of $\mathscr{C}_1$ is approximately _____ Blos.

    b. (1) $m_G(\mathscr{C}_2) \doteq $ _____; therefore,
        (2) the perimeter of $\mathscr{C}_2$ is approximately _____ .
        (3) $m_B(\mathscr{C}_2) \doteq $ _____; therefore,
        (4) the perimeter of $\mathscr{C}_2$ is approximately _____ .

2. Suppose $\mathscr{C}$ is a union-of-line-segments curve; that is, $\mathscr{C} = \overline{AB} \cup \overline{BC} \cup \overline{CD} \cup \overline{DE} \cup \overline{EF}$. Complete the accompanying tables.

a.

Line segment	Approximate measure	g.p.e	r.e.
$\overline{AB}$	$^+3.5$		
$\overline{BC}$	$^+6.7$		
$\overline{CD}$	$^+12.4$		
$\overline{DE}$	$^+8.3$		
$\overline{EF}$	$^+9.2$		
$\mathscr{C}$			

b.

Line segment	Approximate measure	g.p.e	r.e.
$\overline{AB}$	$^+3.54$		
$\overline{BC}$	$^+6.70$		
$\overline{CD}$	$^+12.42$		
$\overline{DE}$	$^+8.35$		
$\overline{EF}$	$^+9.21$		
$\mathscr{C}$			

3. If $\mathscr{C} = \overline{AB} \cup \overline{CD}$ is a union-of-line-segments curve, $m(\overline{AB}) \doteq x$ with g.p.e. $= a$ and $m(\overline{CD}) \doteq y$ with g.p.e. $= b$, determine the r.e. of these approximations and determine the r.e. of the approximation, $m(\mathscr{C}) \doteq x + y$.

congruence and similarity

6.1 CLASSIFICATION OF POLYGONS

Recall that a *polygon* is a simple closed curve which is the union of line segments; the line segments whose union form a polygon are called *sides* of the polygon, the endpoints of the sides are called *vertices*. Each pair of sides with a common endpoint is a subset of an angle having this common point as vertex; every such angle is an *angle of the polygon*. Two sides of a polygon are *adjacent sides* if and only if they have a common endpoint. Two vertices of a polygon are *consecutive vertices* if and only if they are endpoints of the same side. Every line segment determined by a pair of nonconsecutive vertices of a polygon is a *diagonal* of the polygon. A polygon is called an *equilateral* polygon if and only if all its sides are, respectively, congruent. A polygon is called a *regular* polygon if and only if all its sides and all its angles are, respectively, congruent.

Many of these ideas are illustrated in Example 6.1.

EXAMPLE 6.1 Consider the polygon suggested by the accompanying model. The sides of the polygon are $\overline{AB}$, $\overline{BC}$, $\overline{CD}$, $\overline{DE}$, $\overline{EF}$, $\overline{FG}$, and $\overline{GA}$. Two adjacent sides are $\overline{GF}$ and $\overline{FE}$. Which sides are adjacent to $\overline{AG}$? to $\overline{CD}$? Vertices A and G are consecu-

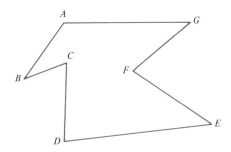

tive, while C and F are nonconsecutive. Are A and F consecutive vertices? Two diagonals are $\overline{CF}$ and $\overline{AF}$. Can you give the remaining diagonals? Since $\overline{BC} \cup \overline{CD}$ is a subset of $\angle BCD$, $\angle BCD$ is an angle of the polygon. Is this polygon a regular polygon? A convex polygon?

To illustrate some additional vocabulary which is used in discussion of polygons we refer to the model in Example 6.1. An angle of a polygon is said to be *included* by the sides of the polygon which are subsets of the sides of the angle. In the polygon of of Example 6.1, $\angle G$ (that is, $\angle AGF$) is included by sides $\overline{AG}$ and $\overline{GF}$. A side of a polygon is said to be *included* by the two angles whose vertices are the endpoints of the side. In the polygon of Example 6.1, side $\overline{AB}$ is included by $\angle B$ and $\angle A$.

6.1.1 Triangles and their classification

A *triangle* is a polygon which is the union of three line segments. One way triangles are classified is according to the number of sides which are congruent. We will see in a later section that the number of congruent sides of a triangle is equal to the number of congruent angles.

A triangle is *scalene* if and only if it has zero sides congruent.
A triangle is *isosceles* if and only if it has at least two sides congruent.
A triangle is *equilateral* if and only if all three sides of the triangle are congruent.

That it would be possible to define scalene, isosceles, and equilateral triangles in terms of angles rather than sides is illustrated in Theorem 6.2, Section 6.3.1 (p. 152).

To illustrate how some of the vocabulary introduced for polygons in general applies to triangles and to introduce some additional vocabulary for triangles we present Example 6.2.

EXAMPLE 6.2 Consider the accompanying model of $\triangle ABC$.

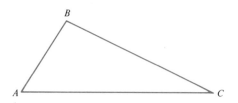

a. The *sides* of $\triangle ABC$ are

   $\overline{AB}$, $\overline{BC}$, and $\overline{CA}$.

   That is, $\triangle ABC = \overline{AB} \cup \overline{BC} \cup \overline{CA}$.

b. The *angles* of $\triangle ABC$ are

   $\overrightarrow{AB} \cup \overrightarrow{AC} = \angle BAC$

   $\qquad\qquad = \angle A$,

   $\overrightarrow{CA} \cup \overrightarrow{CB} = \angle ACB$

   $\qquad\qquad = \angle C$,

   $\overrightarrow{BA} \cup \overrightarrow{BC} = \angle ABC$

   $\qquad\qquad = \angle B$.

c. The *vertices* of $\triangle ABC$ are:

   A, B, and C.

d. $\angle B$ is included by sides $\overline{CB}$ and $\overline{BA}$.

e. Side $\overline{AB}$ is included by $\angle A$ and $\angle B$.

f. $\angle A$ is said to be the *angle opposite* $\overline{BC}$.

g. $\overline{BC}$ is said to be the *side opposite* $\angle A$.

h. Give the sides which are opposite $\angle C$ and $\angle B$, respectively, and the angles opposite $\overline{AB}$ and $\overline{AC}$, respectively.

Exercise

Draw a Venn diagram to show how the sets of equilateral, scalene, and isosceles triangles, and polygons are related.

6.1.2 Quadrilaterals and their classification

A *quadrilateral* is a polygon which is the union of four line segments. Two sides of a quadrilateral are *opposite sides* if and only if they have no points in common. Two vertices of a quadrilateral which are not endpoints of the same side are *opposite vertices*. The line segment determined by a pair of opposite vertices is a *diagonal* of the quadrilateral.

EXAMPLE 6.3 Consider the accompanying model of a quadrilateral with vertices A, B, C, and D.

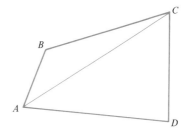

a. $\overline{AB}$ and $\overline{CD}$ are opposite sides.
b. What is the other pair of opposite sides?
c. $\overline{AB}$ and $\overline{AD}$ are adjacent sides.
d. What other side is adjacent to $\overline{AB}$?
e. What sides are adjacent to $\overline{CD}$?
f. $\overline{AC}$ is a diagonal.
g. What is another diagonal?

We use some of the vocabulary introduced above to classify quadrilaterals.

A *trapezoid* is a quadrilateral which has exactly one pair of opposite sides parallel.
A *parallelogram* is a quadrilateral which has a pair of opposite sides congruent and parallel.
A *rhombus* is a parallelogram which has all four sides congruent.
A *rectangle* is a parallelogram which has a right angle.
A *square* is a rectangle which has all four sides congruent.

Since the four vertices of a quadrilateral determine the four line segments whose union is the quadrilateral, we think of every four

points, no three of which are collinear, as determining a quadrilateral. If a quadrilateral is a parallelogram, rectangle, square, or
trapezoid determined by points *A*, *B*, *C*, and *D*, we will denote it
by ⊏*ABCD*, ▭*ABCD*, ▢*ABCD*, and ⌓*ABCD*, respectively.

Exercise set 6.1

1. Draw a Venn diagram to show how the sets of squares, rectangles, parallelograms, rhombuses, trapezoids, quadrilaterals,
and all polygons are related.

2. What types of triangle are regular polygons?

3. What types of quadrilateral are regular polygons?

4. Define each of the following:

   a. pentagon   **b.** hexagon   **c.** octagon
   d. isosceles trapezoid.

5. Consider the polygon suggested by the accompanying model.

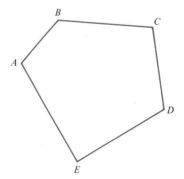

   a. Give the sides of the polygon.
   b. Give the vertices which are consecutive to *C*.
   c. Give the angles of the polygon.
   d. Is the polygon a convex curve?   Explain your answer.
   e. Is $\overline{AC}$ a diagonal?   Explain your answer.
   f. Give at least four diagonals of the polygon.
   g. How many diagonals does the polygon have?
   h. Which of the words you defined in Exercise 4 apply to the
   polygon?

6. Answer questions a–h of Exercise 5 for the polygon suggested by the accompanying model.

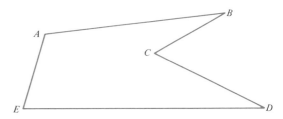

7. It is easy to observe that a convex polygon with three sides has zero diagonals, and as in the accompanying illustration, that a convex polygon with four sides has two diagonals. Label

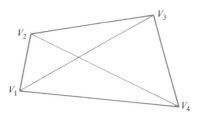

the vertices of a convex polygon with n sides by V_1, V_2, V_3, . . . , V_n, then complete the following tables—you may need to draw models until you observe a pattern in the numbers in the table.

a.
(For a triangle) vertices	V_1	V_2	V_3	Total
Number of new diagonals from:	0	0	0	0

b.
(For a quadrilateral) vertices	V_1	V_2	V_3	V_4	Total
Number of new diagonals from:	1	1	0	0	2

c.
(For a pentagon) vertices	V_1	V_2	V_3	V_4	V_5	Total
Number of new diagonals from:	2	2	?	?	?	?

d.
(For a hexagon) vertices	V_1	V_2	V_3	V_4	V_5	V_6	Total
Number of new diagonals from:	3	?	?	?	?	?	?

e. Complete similar tables for a 7-gon (a polygon with seven sides), an 8-gon, and a 9-gon.

f. Summarize the information (and go beyond) by completing the following table:

Number of sides	3	4	5	6	7	8	9	10	11
Number of diagonals									

g. From the pattern in part f, try to write a formula which gives the number of diagonals of a polygon with n sides.

h. Does the number of diagonals for each entry in the table in part f agree with $\dfrac{n(n-3)}{2}$, where n is the number of sides?

i. Use the expression $\dfrac{n(n-3)}{2}$ to determine the number of diagonals of a 15-gon; of a 27-gon.

6.2 CONGRUENT TRIANGLES

By our definition, a triangle is a polygon which is the union of three line segments; since the endpoints of a line segment determine the line segment, every three noncollinear points determine exactly three line segments whose union is a triangle. Thus we can think of every three noncollinear points as determining a triangle. If A, B, and C are three noncollinear points, we denote the triangle determined by them as $\triangle ABC$.

To turn our attention to the concept of congruence for triangles, let us consider $\triangle ABC$ and $\triangle DEF$. The definition of congruent triangles which we will give uses a one-to-one correspondence between the vertices of the two triangles. Since each triangle has three vertices there are six possible one-to-one correspondences between the vertices. For $\triangle ABC$ and $\triangle DEF$ the following are the six possible one-to-one correspondences between the vertices.

(1)	(2)	(3)	(4)	(5)	(6)
$A \leftrightarrow D$	$A \leftrightarrow D$	$A \leftrightarrow E$	$A \leftrightarrow E$	$A \leftrightarrow F$	$A \leftrightarrow F$
$B \leftrightarrow E$	$B \leftrightarrow F$	$B \leftrightarrow D$	$B \leftrightarrow F$	$B \leftrightarrow D$	$B \leftrightarrow E$
$C \leftrightarrow F$	$C \leftrightarrow E$	$C \leftrightarrow F$	$C \leftrightarrow D$	$C \leftrightarrow E$	$C \leftrightarrow D.$

We will use a shorthand notation to indicate these one-to-one correspondences as follows: $ABC \leftrightarrow DEF$, $ABC \leftrightarrow DFE$, $ABC \leftrightarrow EDF$, $ABC \leftrightarrow EFD$, $ABC \leftrightarrow FDE$, and $ABC \leftrightarrow FED$ denote the one-to-one correspondences (1)–(6), respectively.

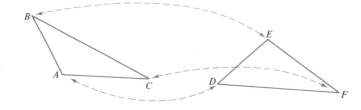

FIG. 6.1
A model to suggest a one-to-one correspondence between the vertices of $\triangle ABC$ and $\triangle DEF$.

Refer to Fig. 6.1 and consider the one-to-one correspondence between the vertices of $\triangle ABC$ and $\triangle DEF$ denoted by $ABC \leftrightarrow DEF$. The correspondence between the vertices of $\triangle ABC$ and $\triangle DEF$ suggests a correspondence between sides of the two triangles,

$$\overline{AB} \leftrightarrow \overline{DE}, \qquad \overline{AC} \leftrightarrow \overline{DF}, \qquad \text{and} \qquad \overline{BC} \leftrightarrow \overline{EF},$$

and a correspondence between the angles,

$$\angle A \leftrightarrow \angle D, \qquad \angle B \leftrightarrow \angle E, \qquad \text{and} \qquad \angle C \leftrightarrow \angle F.$$

With respect to the one-to-one correspondence $ABC \leftrightarrow DEF$ between the vertices of $\triangle ABC$ and $\triangle DEF$, the paired sides and angles as indicated above are called *corresponding sides* and *corresponding angles*, respectively.

The reader will readily agree that the triangles suggested by the models in Fig. 6.1 are not congruent. What then, in addition to a one-to-one correspondence between the vertices, is necessary for triangles to be congruent? This is suggested in Example 6.4 and stated formally in Definition 6.1.

EXAMPLE 6.4 Consider $\triangle GFE$ and $\triangle JIH$ suggested by the accompanying models. With respect to the one-to-one correspondence $GFE \leftrightarrow JIH$, the following are apparently true.

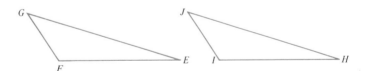

$$\overline{GF} \cong \overline{JI}, \qquad \overline{FE} \cong \overline{IH}, \qquad \overline{GE} \cong \overline{JH}$$

and

$$\angle G \cong \angle J, \qquad \angle F \cong \angle I, \qquad \angle E \cong \angle H.$$

That is, with respect to the one-to-one correspondence between the vertices, corresponding sides and angles are, respectively, congruent.

The remarks in Example 6.4 suggest Definition 6.1.

Definition 6.1 For all triangles ABC and DEF, a one-to-one correspondence between the vertices is a *congruence* between the triangles if and only if corresponding sides and corresponding angles with respect to the one-to-one correspondence are respectively congruent. If $ABC \leftrightarrow DEF$ is a congruence between $\triangle ABC$ and $\triangle DEF$, we say the triangles are *congruent* under the one-to-one correspondence and write $\triangle ABC \cong \triangle DEF$.

A comment on the notation in Definition 6.1 is in order. When we write

$\triangle ABC \cong \triangle DEF$, the one-to-one correspondence between the vertices is $ABC \leftrightarrow DEF$;
$\triangle BAC \cong \triangle DEF$, the one-to-one correspondence between the vertices is $BAC \leftrightarrow DEF$.

Therefore, to write $\triangle ABC \cong \triangle DEF$ for example, means that corresponding sides are

$\overline{AB}$ and $\overline{DE}$, $\qquad \overline{AC}$ and $\overline{DF}$, $\qquad$ and $\qquad \overline{BC}$ and $\overline{EF}$,

and corresponding angles are

$\angle A$ and $\angle D$, $\qquad \angle B$ and $\angle E$, $\qquad$ and $\qquad \angle C$ and $\angle F$.

That is, $\triangle ABC \cong \triangle DEF$ if and only if

$\overline{AB} \cong \overline{DE}$, $\qquad \overline{AC} \cong \overline{DF}$, $\qquad$ and $\qquad \overline{BC} \cong \overline{EF}$

and

$\angle A \cong \angle D$, $\qquad \angle B \cong \angle E$, $\qquad$ and $\qquad \angle C \cong \angle F$.

Theorem 6.1 The relation $\cong$ on the set of all triangles has the following properties:

a. For every triangle ABC, $\triangle ABC \cong \triangle ABC$.
b. For all triangles ABC and DEF, if $\triangle ABC \cong \triangle DEF$, then $\triangle DEF \cong \triangle ABC$.
c. For all triangles ABC, DEF, and GHI, if $\triangle ABC \cong \triangle DEF$ and $\triangle DEF \cong \triangle GHI$, then $\triangle ABC \cong \triangle GHI$.

The proof is left to the reader.

Exercise set 6.2

1. Consider $\triangle ABC$ and $\triangle DEF$.

    a. $ABC \leftrightarrow DEF$ is a one-to-one correspondence between the vertices of the two triangles. With respect to this one-to-one correspondence:

(1) $\overline{AB}$ and _____ are corresponding sides.

(2) _____ and $\overline{DF}$ are corresponding sides.

(3) _____ and _____ is the third pair of corresponding sides.

(4) $\angle A$ and _____ are corresponding angles.

(5) _____ and $\angle E$ are corresponding angles.

(6) _____ and _____ is the third pair of corresponding angles.

b. $BAC \leftrightarrow DEF$ indicates another one-to-one correspondence between the vertices of $\triangle ABC$ and $\triangle DEF$. Draw a model (see Fig. 6.1) which suggests the one-to-one correspondence. Two corresponding sides are $\overline{BA}$ and $\overline{DE}$. Indicate the other two pairs of corresponding sides and the three pairs of corresponding angles determined by this one-to-one correspondence.

c. Indicate the four remaining one-to-one correspondences between the vertices of $\triangle ABC$ and $\triangle DEF$ and for each draw a model (see Fig. 6.1) which suggests the one-to-one correspondence.

2. For triangle ABC there are six one-to-one correspondences of the set $\{A,B,C\}$ of vertices of $\triangle ABC$ with itself: $ABC \leftrightarrow ABC$ and $ABC \leftrightarrow BAC$ are two of them. List the remaining one-to-one correspondences between the vertices and list the corresponding sides and angles with respect to all six one-to-one correspondences.

3. Prove Theorem 6.1.

6.3 MINIMUM CONDITIONS FOR TRIANGLE CONGRUENCE

The definition of congruence of triangles requires that *three* sides and *three* angles of one are respectively congruent to *three* sides and *three* angles of the other. That is, according to the definition, in order to show that two triangles are congruent it is necessary to show a total of six (line segment and angle) congruences. A natural question is to ask whether the number of such congruences can be reduced and if so, what the minimum number is. The reader should consider Example 6.5 and the exercises in Exercise set 6.3, which suggest the answer to this question.

EXAMPLE 6.5 We use the accompanying models to suggest whether △*ABC* ≅ △*DEF* under the one-to-one correspondence *ABC* ↔ *DEF* when

$$\overline{AB} \cong \overline{DE}, \qquad \overline{BC} \cong \overline{EF}, \qquad \text{and} \qquad \angle C \cong \angle F.$$

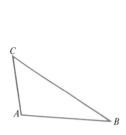 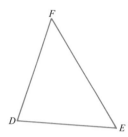 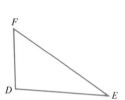

We leave to the reader the problem of convincing himself that both of the triangles labeled *DEF* satisfy the conditions of

$$\overline{AB} \cong \overline{DE}, \qquad \overline{BC} \cong \overline{EF}, \qquad \text{and} \qquad \angle C \cong \angle F,$$

but that only one of the models suggests a triangle which is congruent to △*ABC*.

Exercise set 6.3

1. Draw models of triangles to suggest whether △*ABC* and △*DEF* are congruent under the one-to-one correspondence *ABC* ↔ *DEF* when:

   A. a. Only ∠*A* ≅ ∠*D*.
   b. Only ∠*A* ≅ ∠*D* and ∠*B* ≅ ∠*E*.
   c. ∠*A* ≅ ∠*D*, ∠*B* ≅ ∠*E*, and ∠*C* ≅ ∠*F*.
   B. a. Only $\overline{AB} \cong \overline{DE}$.
   b. Only $\overline{AB} \cong \overline{DE}$, and $\overline{BC} \cong \overline{EF}$.
   c. $\overline{AB} \cong \overline{DE}$, $\overline{BC} \cong \overline{EF}$, and $\overline{AC} \cong \overline{DF}$.
   C. a. Only ∠*A* ≅ ∠*D* and $\overline{AB} \cong \overline{DE}$.
   b. Only ∠*A* ≅ ∠*D*, $\overline{AB} \cong \overline{DE}$, and $\overline{BC} \cong \overline{EF}$.
   c. Only ∠*A* ≅ ∠*D*, $\overline{AB} \cong \overline{DE}$, and $\overline{AC} \cong \overline{DF}$.
   D. a. Only $\overline{AB} \cong \overline{DE}$, ∠*A* ≅ ∠*D*, and ∠*C* ≅ ∠*F*.
   b. Only $\overline{AB} \cong \overline{DE}$, ∠*A* ≅ ∠*D*, and ∠*B* ≅ ∠*E*.

2. State the generalizations which are suggested by Exercise 1.

   The minimum number of conditions for two triangles to be congruent which are suggested by Exercise 1 above are provable as theorems; however, our development is not sufficiently complete for these proofs, so we present them as postulates.

Postulate 6.1 (SSS triangle congruence postulate) A one-to-one correspondence between the vertices of two triangles is a congruence if and only if the three pairs of corresponding sides are respectively congruent.

Postulate 6.2 (SAS triangle congruence postulate) A one-to-one correspondence between the vertices of two triangles is a congruence if and only if two sides and their included angle of one triangle are respectively congruent to the corresponding sides and included angle of the second.

Postulate 6.3 (ASA triangle congruence postulate) A one-to-one correspondence between the vertices of two triangles is a congruence if and only if two angles and their included side of one triangle are respectively congruent to the corresponding angles and included side of the second.

Postulate 6.4 (SAA triangle congruence postulate) A one-to-one correspondence between the vertices of two triangles is a congruence if and only if one side and two angles of one triangle are respectively congruent to the corresponding side and angles of the second.

Exercise set 6.4

1. Consider $\triangle ABC$, $\triangle DEF$, $\triangle GHI$, and $\triangle JKL$ with measures of sides and angles as given in the following tabulation:

$\triangle ABC$	$\triangle DEF$	$\triangle GHI$	$\triangle JKL$
$m(\overline{AB}) = {}^+4$	$m_\circ(\angle F) = {}^+40$	$m_\circ(\angle G) = {}^+30$	$m(\overline{JK}) = {}^+4$
$m(\overline{AC}) = {}^+5$	$m(\overline{DF}) = {}^+5$	$m_\circ(\angle I) = {}^+40$	$m(\overline{KL}) = {}^+5$
$m_\circ(\angle A) = {}^+30$	$m(\overline{EF}) = {}^+4$	$m(\overline{GI}) = {}^+4$	$m(\overline{JL}) = {}^+7$
$m_\circ(\angle B) = {}^+40$			

For each pair of triangles given in parts a–g on the next page:
(i) Indicate whether they are congruent.
(ii) If they are congruent, indicate the required one-to-one correspondence between the vertices and state the tri-

angle congruence postulate which justifies that they are congruent.

a. $\triangle ABC$ and $\triangle DEF$ b. $\triangle ABC$ and $\triangle GHI$
c. $\triangle DEF$ and $\triangle GHI$ d. $\triangle ABC$ and $\triangle JKL$
e. $\triangle DEF$ and $\triangle JKL$ f. $\triangle GHI$ and $\triangle JKL$
g. $\triangle ABC$ and $\triangle ABC$

2. If $\triangle ABC \cong \triangle DEF$, is it true in general that $\triangle ABC \cong \triangle EDF$? Describe the conditions necessary so that $\triangle ABC \cong \triangle EDF$.

3. Consider the accompanying model.

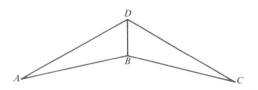

$\overline{AB} \cong \overline{BC}, \qquad \overline{AD} \cong \overline{DC}.$

a. Is $\triangle ABD \cong \triangle CBD$?
b. If yes, indicate the one-to-one correspondence between the vertices which is a congruence and state the triangle congruent postulate which justifies that the triangles are congruent.

4. The accompanying model is of a rectangle with diagonal $\overline{EG}$.

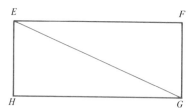

a. Is $\triangle HEG \cong \triangle FGE$?
b. Answer Exercise 3b about $\triangle HEG$ and $\triangle FGE$.

5. The accompanying model is of an isosceles trapezoid with diagonal $\overline{LJ}$.

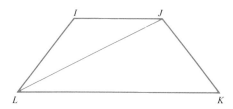

   a. Is $\triangle LIJ \cong \triangle JKL$?

   b. Answer Exercise 3b about $\triangle LIJ$ and $\triangle JKL$.

6. The accompanying model is of an isosceles trapezoid with diagonals $\overline{MO}$ and $\overline{NP}$, $\overline{MO} \cong \overline{NP}$.

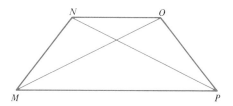

   a. Is $\triangle MNO \cong \triangle PON$?

   b. Answer Exercise 3b about $\triangle MNO$ and $\triangle PON$.

7. In the accompanying model $\overleftrightarrow{QR} \parallel \overleftrightarrow{TS}$ and $\angle QTR \cong \angle SRT$.

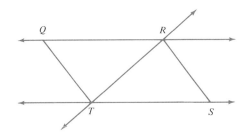

   a. Is $\triangle QTR \cong \triangle SRT$?

   b. Answer Exercise 3b about $\triangle QTR$ and $\triangle SRT$.

8. The model on the next page is of an isosceles trapezoid such that $\overline{UZ} \cong \overline{YX}$ and $\overline{VZ} \cong \overline{WY}$. Show that $\triangle UVZ \cong \triangle XWY$

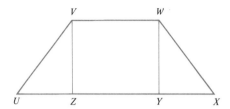

6.3.1   Some applications of the triangle congruence postulates

We indicated in Section 6.1.1 that it would be possible to define scalene, isosceles, and equilateral triangles in terms of angles rather than sides; we illustrate for the case of isosceles triangles in Theorem 6.2.

Theorem 6.2   For every $\triangle ABC$, $\triangle ABC$ is isosceles if and only if two angles of $\triangle ABC$ are congruent.

A.   Prove: If $\triangle ABC$ is isosceles, then two angles are congruent.

PROOF

1. $\triangle ABC$ is isosceles.	1. Given
2. $\therefore$ There are two sides of $\triangle ABC$ which are congruent, say, $\overline{AC} \cong \overline{BC}$.	2. Definition of isosceles triangle
3. $ACB \leftrightarrow BCA$ is a one-to-one correspondence of the set $\{A,B,C\}$ of vertices of $\triangle ABC$ with itself.	3. Definition of one-to-one correspondence
4. $\angle C \cong \angle C$.	4. Reflexive property of $\cong$ for angles
5. $\overline{BC} \cong \overline{CB}$.	5. Exercise 4 of Exercise Set 4.3, p. 94, Step 2
6. $\therefore \overline{AC} \cong \overline{CB}$.	6. Transitive property of $\cong$ for line segments
7. $\overline{CA} \cong \overline{AC}$.	7. Exercise 4 of Exercise Set 4.3, p. 94
8. $\therefore \overline{CA} \cong \overline{CB}$.	8. Transitive property of $\cong$ for line segments

9. ∴ $\overline{CB} \cong \overline{CA}$.

9. Symmetric property of ≅ for line segments

10. ∴ △$ACB \cong$ △BCA.

10. SAS triangle congruence postulate

11. ∴ ∠$A \cong$ ∠B.

11. Definition of ≅ for triangles

B. Prove: If two angles of △ABC are congruent, then △ABC is isosceles.

PROOF

12. Two angles of △ABC are congruent, say, ∠$A \cong$ ∠B.

12. Given

13. ∴ ∠$B \cong$ ∠A.

13. Why?

14. $\overline{AB} \cong \overline{AB}$.

14. ?

15. $\overline{AB} \cong \overline{BA}$.

15. ?

16. ∴ △$ABC \cong$ △BAC.

16. ?

17. ∴ $\overline{AC} \cong \overline{BC}$.

17. ?

18. ∴ △ABC is isosceles.

18. ?

Using triangle congruence postulates, many relationships between "parts" of quadrilaterals and triangles can be proved, we illustrate in Proposition 6.1.

Proposition 6.1 Each diagonal of a parallelogram $ABCD$ forms with the sides of ▱$ABCD$ two congruent triangles. (Reference to the accompanying model will help the reader follow the proof

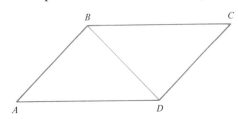

that the diagonal $\overline{BD}$ forms with parallelogram $ABCD$ the congruent triangles ABD and CDB.)

PROOF

1. Points $A, B, C,$ and D determine ▱$ABCD$.

1. Given

2. ∴ $\overline{AB} \parallel \overline{CD}$ and $\overline{AB} \cong \overline{CD}$ or $\overline{BC} \parallel \overline{AD}$ and $\overline{BC} \cong \overline{AD}$.

2. Definition of parallelogram

3. Suppose $\overline{AB} \parallel \overline{CD}$ and $\overline{AB} \cong \overline{CD}$.   3. Assumption
4. $\therefore \overleftrightarrow{AB} \parallel \overleftrightarrow{CD}$.   4. Why?
5. $\overleftrightarrow{BD}$ is a transversal for $\overleftrightarrow{AB}$ and $\overleftrightarrow{CD}$.   5. ?
6. $\angle ABD$ and $\angle CDB$ are alternate interior   6. ?
    angles.
7. $\therefore \angle ABD \cong \angle CDB$.   7. ?
8. $\overline{BD} \cong \overline{BD}$.   8. ?
9. $\therefore \triangle ABD \cong \triangle CDB$.   9. ?

We now leave to the reader the problems of showing that $\triangle ABD \cong \triangle CDB$ under the assumption that $\overline{BC} \parallel \overline{AD}$ and $\overline{BC} \cong \overline{AD}$ and that diagonal $\overline{AC}$ forms with the sides of $\square ABCD$ the congruent triangles ABC and CDA.

Exercise set 6.5

1. Prove: If two sides of a triangle are congruent, then the opposite angles of the two sides are congruent.

2. Prove: If a triangle is scalene, then it has no two angles which are congruent.

3. Prove: If a triangle is equilateral, then it is equiangular. (That is, all three angles are congruent.)

4. Prove: If two adjacent sides of a parallelogram are congruent, then the parallelogram is a rhombus.

5. Prove: Each pair of opposite sides of a parallelogram is respectively congruent and parallel.

6. Prove: Every rectangle has four right angles.

7. Prove: The diagonals of a parallelogram bisect each other.

8. Prove: If the diagonals of a parallelogram are perpendicular, then the parallelogram is a rhombus.

9. Prove: The diagonals of a rectangle are congruent.

10. Prove: If a rhombus has a right angle, then it is a square.

11. State a definition of a square using the concept of a rhombus.

6.4 CONGRUENCE OF POLYGONS, CIRCLES, AND CONGRUENCE IN GENERAL

Congruence of convex polygons is a natural generalization of congruence of triangles.

Definition 6.2 For all convex polygons $\mathscr{P}_1$ and $\mathscr{P}_2$, a one-to-one correspondence between the vertices of $\mathscr{P}_1$ and $\mathscr{P}_2$ is a *congruence* between the polygons if and only if corresponding sides and angles, with respect to the one-to-one correspondence, are respectively congruent. Two polygons are *congruent* if and only if there exists a congruence between them.

Definition 6.2 is illustrated in Examples 6.6 and 6.7.

EXAMPLE 6.6 The two convex polygons suggested in the accompanying models are clearly not congruent because there is not a one-to-one correspondence between the vertices.

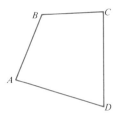
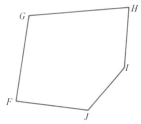

EXAMPLE 6.7 In order to show that the convex quadrilaterals suggested by the accompanying models are congruent, it would be necessary to show that one of the 24 (that is, $4 \cdot 3 \cdot 2 \cdot 1$) one-to-one correspondences between their vertices is a congruence.

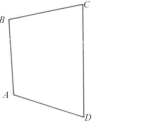
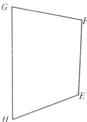

The models suggest that $ABCD \leftrightarrow EFGH$ is the required one-to-one correspondence.

Definition 6.3 For all circles $\mathscr{C}_1$ and $\mathscr{C}_2$, $\mathscr{C}_1$ is congruent to $\mathscr{C}_2$ if and only if the radius of $\mathscr{C}_1$ is equal to the radius of $\mathscr{C}_2$.

From a purely mathematical point of view it would be desirable to give a definition of congruence for geometric figures in general and then from this definition to prove as theorems the several results which specify when two line segments, two angles, two triangles, and so on, are congruent. Such an approach is possible; the unfortunate thing is that the general definition, although seemingly very simple, is difficult to apply to obtain the specific criterion for determining whether two particular geometric figures are congruent. For the sake of completeness we give the general definition.

Definition 6.4 For all $\mathscr{A}$ and $\mathscr{B}$ which are subsets of S, a congruence between $\mathscr{A}$ and $\mathscr{B}$ is a one-to-one correspondence between $\mathscr{A}$ and $\mathscr{B}$ which preserves distance.

6.5 SIMILARITY

The relation of congruence on geometric figures is the mathematical abstraction of the physical concepts of "same size" and "same shape." Another important relation on geometric figures is the mathematical abstraction of "same shape" only; this relation is called *similarity*. We will discuss this relation especially for triangles, since this is the most common application of this relation, and generalize from this to other geometric figures.

Consider the model triangles ABC, DEF, and GHI in Fig. 6.2. Do you consider the models of $\triangle ABC$ and $\triangle GHI$ to have the same

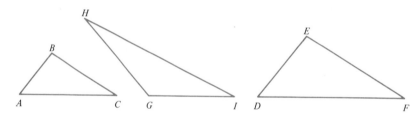

FIG. 6.2
Models to suggest "same shape" for triangles.

shape? How about the models of $\triangle ABC$ and $\triangle DEF$? How do the angles—$\angle A$ and $\angle D$, $\angle B$ and $\angle E$, and $\angle C$ and $\angle F$—appear to be related? Do the models seem to suggest that $\angle A \cong \angle D$, $\angle B \cong \angle E$, and $\angle C \cong \angle F$? Do you observe any relation between the sides of $\triangle ABC$ and $\triangle DEF$? Approximate the measures of the sides of $\triangle ABC$ and $\triangle DEF$ and compare $\dfrac{m(\overline{AB})}{m(\overline{DE})}$ to $\dfrac{m(\overline{BC})}{m(\overline{EF})}$ and to $\dfrac{m(\overline{AC})}{m(\overline{DF})}$. The questions and observations above should suggest Definition 6.5.

Definition 6.5 For all triangles ABC and DEF, the one-to-one correspondence $ABC \leftrightarrow DEF$ between the vertices of $\triangle ABC$ and $\triangle DEF$ is a *similarity* if and only if corresponding angles are congruent and the measures of corresponding sides are proportional.

Whenever a similarity exists between two triangles ABC and DEF, we say they are *similar* and write $\triangle ABC \sim \triangle DEF$.

Theorem 6.3 The relation $\sim$ on triangles has the following properties:

1. For every triangle ABC, $\triangle ABC \sim \triangle ABC$.
2. For all triangles ABC and DEF, if $\triangle ABC \sim \triangle DEF$, then $\triangle DEF \sim \triangle ABC$.
3. For all triangles ABC, DEF, and GHI, if $\triangle ABC \sim \triangle DEF$ and $\triangle DEF \sim \triangle GHI$, then $\triangle ABC \sim \triangle GHI$.

That is, $\sim$ is an equivalence relation on the set of all triangles.

The definition of the relation of congruence on triangles (Definition 6.1, p. 145) mentions six parts of the two triangles, as does Definition 6.5, the definition of the relation of similarity; thus, it is very natural to question whether the number of corresponding parts needed to determine a similarity can be decreased and what the minimum number is. It also seems natural to question whether we have analogs to the SSS, SAS, ASA, and SAA conditions for congruence; some experimentation with models of triangles will suggest an affirmative answer to the question. We state the conditions as postulates, although a more formal and complete development of geometry would allow for proofs of these statements.

Postulate 6.5 (*SSS triangle similarity postulate*) A one-to-one correspondence between the vertices of two triangles is a similarity if and only if the measures of the corresponding sides are proportional.

Postulate 6.6 (*SAS triangle similarity postulate*) A one-to-one correspondence between the vertices of two triangles is a similarity if and only if the measures of two pairs of corresponding sides are proportional and their included angles are congruent.

Postulate 6.7 (*AA triangle similarity postulate*) A one-to-one correspondence between the vertices of two triangles is a similarity if and only if two angles of one triangle are congruent to the corresponding two angles of the other.

For the sake of completeness we give a general definition of the similarity relation defined for all geometric figures.

Definition 6.6 For all $\mathscr{F}_1$ and $\mathscr{F}_2$ which are geometric figures, a one-to-one correspondence between $\mathscr{F}_1$ and $\mathscr{F}_2$ is a *similarity* if and only if for all points $X, Y \in \mathscr{F}_1$, which are paired with points X' and Y' in $\mathscr{F}_2$, $XY = k \times (X'Y')$, $k \in \mathbb{R}$, $k > 0$. If a similarity exists between $\mathscr{F}_1$ and $\mathscr{F}_2$, we say they are *similar* and write $\mathscr{F}_1 \sim \mathscr{F}_2$.

Exercise set 6.6

1. Consider $\triangle ABC$, $\triangle DEF$, $\triangle GHI$, $\triangle JKL$, $\triangle MNO$, $\triangle PQR$, and $\triangle STU$ with measures of sides and angles as given in the accompanying tabulation. For each pair of triangles given in parts a–g:
   (i) Indicate whether they are similar.
   (ii) If they are similar, indicate the required one-to-one correspondence between the vertices and state the triangle similarity postulate which justifies that they are similar.

   a. $\triangle ABC$ and $\triangle DEF$ b. $\triangle ABC$ and $\triangle GHI$
   c. $\triangle ABC$ and $\triangle JKL$ d. $\triangle ABC$ and $\triangle PQR$
   e. $\triangle ABC$ and $\triangle STU$ f. $\triangle DEF$ and $\triangle JKL$
   g. $\triangle DEF$ and $\triangle MNO$.

△ABC	△DEF	△GHI	△JKL
$m(\overline{AB}) = {}^+4$	$m\circ(\angle F) = {}^+40$	$m\circ(\angle G) = {}^+30$	$m\circ(\angle J) = {}^+40$
$m(\overline{AC}) = {}^+5$	$m(\overline{FD}) = {}^+5$	$m\circ(\angle I) = {}^+40$	$m(\overline{JK}) = {}^+4$
$m\circ(\angle A) = {}^+30$	$m(\overline{EF}) = {}^+4$	$m(\overline{GI}) = {}^+12$	$m(\overline{KL}) = {}^+5$
$m\circ(\angle B) = {}^+40$			

△MNO	△PQR	△STU
$m\circ(\angle O) = {}^+40$	$m(\overline{PQ}) = {}^+2$	$m(\overline{ST}) = {}^+4$
$m(\overline{MO}) = {}^+15$	$m(\overline{PR}) = {}^+2\frac{1}{2}$	$m(\overline{SU}) = {}^+5$
$m(\overline{NO}) = {}^+12$	$m\circ(\angle P) = {}^+30$	$m\circ(\angle S) = {}^+30$

2. Prove: If a line intersects two sides of a triangle forming corresponding line segments whose measures are proportional, then the line is parallel to the third side, as illustrated in the accompanying model.

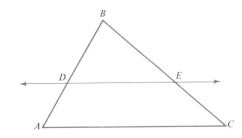

Given: △ABC, $\overleftrightarrow{DE}$, and $\dfrac{m(\overline{DB})}{m(\overline{AD})} = \dfrac{m(\overline{EB})}{m(\overline{CE})}$.

Prove: $\overleftrightarrow{DE} \parallel \overline{AC}$.

(*Hint:* Show $\angle A \cong \angle EDB$.)

3. Prove: The line which bisects one side of a triangle and is parallel to another, bisects the third, as illustrated in the model on the next page.

Given: △ABC, $\overleftrightarrow{DE}$ such that $\overleftrightarrow{DE}$ bisects $\overline{BA}$ and $\overleftrightarrow{DE} \parallel \overline{AC}$.

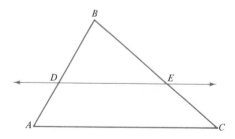

Prove: $\overleftrightarrow{DE}$ bisects $\overline{BC}$.

(*Hint:*   Show $\triangle ABC \sim \triangle DBE$.)

4. **a.**   Prove: The line segment determined by the midpoints of two sides of a triangle is parallel to the third side.

   b.   Prove: The measure of the line segment determined by the midpoints of two sides of a triangle is one-half the measure of the third side.   (See the accompanying model.)

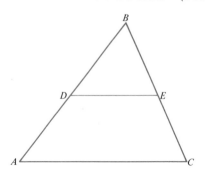

Given: $\triangle ABC, D$ is the midpoint of $\overline{AB}$, and E is the midpoint of $\overline{CB}$.

Prove: **a.** $\overline{DE} \parallel \overline{AC}$.

   b. $m(\overline{DE}) = \dfrac{^{+}1}{^{+}2} \times m(\overline{AC})$.

5.   Indicate which of the figures suggested by the accompanying models belong to the same equivalence class determined by $\sim$.

1

2

3

4

5

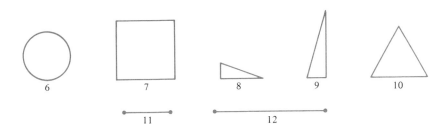

6. Prove that the diagonals of an isosceles trapezoid form with the sides of the trapezoid two pairs of similar triangles as illustrated in the accompanying model.

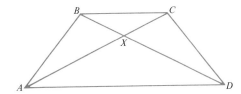

Given: $\square ABCD$ with $\overline{AB} \cong \overline{CD}$.

Prove: $\triangle BCX \sim \triangle DAX$, $\triangle ABX \sim \triangle DCX$.

6.6 STRAIGHTEDGE AND COMPASS CONSTRUCTIONS

In this section we give some theorems and postulates which guarantee the existence of line segments and angles which are congruent to given line segments and angles. From these theorems and postulates, together with the triangle congruence postulates, we will justify some of the elementary classical straightedge and compass constructions that you learned in high school. It may turn out to be of interest to the reader that some rather sophisticated-looking postulates and theorems are necessary to justify these simple constructions.

We present first some (at this point) seemingly unrelated properties of circles.

Theorem 6.4 For every circle $\mathscr{C}$ with center P, $P \in \text{Int}(\mathscr{C})$.

The proof is left as an easy exercise for the reader.

Postulate 6.8 (Two circle postulate) If $\mathcal{C}_1$ and $\mathcal{C}_2$ are two circles with radii a and b, respectively, and with distance d between their centers where a, b, and d are such that each is less than the sum of the other two, then $\mathcal{C}_1$ and $\mathcal{C}_2$ intersect in two points which are on opposite sides of the line containing their centers.

The diagrams in Example 6.8 illustrate Postulate 6.8.

EXAMPLE 6.8 The accompanying models illustrate the various cases of Postulate 6.8.

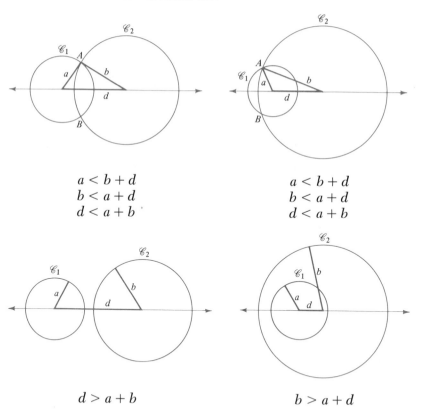

$$a < b + d$$
$$b < a + d$$
$$d < a + b$$

$$a < b + d$$
$$b < a + d$$
$$d < a + b$$

$$d > a + b$$

$$b > a + d$$

It is now possible to prove the following:

Theorem 6.5 For every circle $\mathcal{C}$ with center P and for every point X in the plane containing $\mathcal{C}$, $\overrightarrow{PX} \cap \mathcal{C}$ contains exactly one point.

Corollary 6.5 For every circle $\mathscr{C}$ with center P, and for every point X in the plane containing $\mathscr{C}$, $\overleftrightarrow{PX} \cap \mathscr{C}$ contains exactly two points.

The most elementary type of construction is the problem of constructing a line segment on a ray which is congruent to a given line segment. The more precise mathematical formulation of the problem is given in Theorem 6.6.

Theorem 6.6 (Point plotting theorem) For every ray AX and for every real number r, there is a unique point D on $\overrightarrow{AX}$ such that $m(\overline{AD}) = r$.

OUTLINE OF PROOF

There exists a coordinate system on $\overleftrightarrow{AX}$ such that the coordinate of A is zero and the coordinate of X is a positive real number. Since the coordinate system is a one-to-one correspondence between $\overleftrightarrow{AX}$ and $\mathbb{R}$, there is a point D whose coordinate is r and hence, $m(\overline{AD}) = AD = |r - 0| = r$.

The physical parallel of Theorem 6.6 is Construction I.

Construction I Given the accompanying model of a line segment AB and the accompanying model of a ray CX, using only a

$$A \bullet\!\!-\!\!-\!\!-\!\!-\!\!-\!\!\bullet\, B \qquad C \bullet\!\!-\!\!-\!\!-\!\!-\!\!-\!\!-\!\!-\!\!-\!\!\bullet\!\!\longrightarrow \atop X$$

compass or straightedge, construct a model of a line segment CD on $\overrightarrow{CX}$ such that $\overline{CD} \cong \overline{AB}$.

STEPS IN THE CONSTRUCTION

"Draw a circle" with radius $r = m(\overline{AB})$ with center at C. The

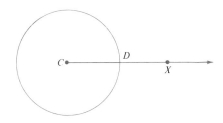

unique point D in the intersection set of the circle and the ray is the required point.

OUTLINE OF PROOF FOR CONSTRUCTION I

By the ruler placement postulate, there is a real number r such that $AB = r$, that is, $m(\overline{AB}) = r$. By Postulate 5.1, there exists a circle with center at C and radius r. It can be proved from Theorem 6.5 and the definition of a circle that this circle intersects $\overrightarrow{CX}$ in exactly one point D. Therefore, $m(\overline{CD}) = r$, and hence $\overline{CD} \cong \overline{AB}$.

Another elementary construction problem is that of constructing a model of an angle which is congruent to a model of a given angle. The precise geometric statement of this problem is given in Theorem 6.7.

Theorem 6.7 (Angle construction theorem) For every ray AB and for every real number r, $0 < r < {}^{+}180$, there exists a unique ray AC in a half-plane determined by $\overleftrightarrow{AB}$ such that $m_{\circ}(\angle CAB) = r$.

The proof of the theorem follows from the ray coordinate postulates in much the same way that Theorem 6.6 follows from the line coordinate postulates.
    The physical parallel of Theorem 6.7 is Construction II.

Construction II Given the accompanying model of an angle ABC and the accompanying model of a ray DX, using a compass

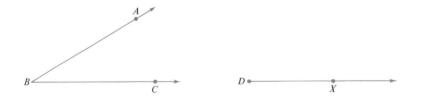

or straightedge, construct a model of an angle YDX such that $\angle YDX \cong \angle ABC$.

STEPS IN THE CONSTRUCTION

"Draw a circle" with center at B and radius r. The circle will intersect $\angle ABC$ in exactly two points Q and P as illustrated in the accompanying model.

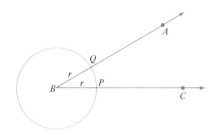

"Draw a circle" with center at D with radius r; $r = m(\overline{BP})$ and $r = m(\overline{BQ})$. The circle will intersect the ray DX in a unique point R as in the accompanying model.

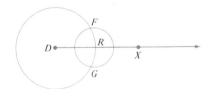

"Draw a circle" with center at R and radius $m(\overline{QP})$. (Use the definition of betweenness and the triangle postulate for the measure of distance to prove that the circles intersect in points F and G as suggested in the model.)

"Draw ray DF."

By our construction, $\overline{DF} \cong \overline{BQ}$, $\overline{RF} \cong \overline{PQ}$, and $\overline{DR} \cong \overline{BP}$; therefore, by the SSS triangle congruence postulate $DRF \leftrightarrow BPQ$ is a congruence between $\triangle DRF$ and $\triangle BPQ$ and hence $\angle B \cong \angle D$.

Exercise set 6.7

1. Another elementary construction which depends on the two circle postulate is to construct a model of a triangle which is congruent to the model of a given triangle. Recall or discover the construction and justify the steps in your construction.

2. For every angle, the bisector of the angle is the ray such that its endpoint is the vertex of the angle; all other points of the

ray are in the interior of the angle, and the ray together with the sides of the angle form two congruent angles.

Recall or discover the construction to determine the bisector of an angle and justify each step in your construction.

3. Recall or discover the construction to determine the midpoint of a line segment.   Prove that your construction is valid.

4. Recall or discover the construction to determine the perpendicular bisector of a line segment.   Prove that your construction is valid.

5. Assume the following:

For every line $\mathscr{L}$ in a plane α and for every point $X \in \alpha$, $X \notin \mathscr{L}$, there exists exactly one line through X perpendicular to $\mathscr{L}$.

Recall or discover the construction of the perpendicular from a point not on a line to the line.   Prove that your construction is valid.

6. Recall or discover the construction to determine the line parallel to a given line $\mathscr{L}$ through a given point P not on $\mathscr{L}$.

7. For all right triangles ABC with $m_\circ(\angle ACB) = {}^{+}90$, $(m(\overline{AC}))^2 + (m(\overline{CB}))^2 = (m(\overline{AB}))^2$.

   a. Determine the measure of the diagonals of a square whose sides have measure ${}^{+}1$.

   b. Construct a line segment whose measure is $\sqrt{{}^{+}2}$; $\sqrt{{}^{+}3}$; ${}^{+}2$; $\sqrt{{}^{+}5}$; $\sqrt{{}^{+}6}$; and so on.

7

---

regions, measure of regions

7.1 DEFINITION OF A REGION

The concepts of simple closed curves and their interiors enable us to give a definition of the concept of a region; this is given in Definition 7.1.

Definition 7.1 A region $\mathscr{R}$ is a set of points in a plane which is the union of a simple closed curve $\mathscr{C}$ and its interior. We denote this region by $\mathrm{Reg}(\mathscr{C})$ (read, "the region determined by $\mathscr{C}$"). The simple closed curve is called the *boundary* of the region it determines.

An illustration is given in Example 7.1

EXAMPLE 7.1 The accompanying models are of a simple closed curve $\mathscr{C}$ and the region determined by $\mathscr{C}$.

We classify regions according to the simple closed curve which "determines the region." Thus the union of a polygon and its interior is a *polygonal* region; the union of a triangle and its interior is a *triangular* region; the union of a circle and its interior is a *circular* region, and so on. We will denote the region determined by $\triangle ABC$ as $\blacktriangle ABC$; similar notation will be used for other regions where appropriate.

EXAMPLE 7.2 Let the triangular region ABC (determined by $\triangle ABC$), line DE, and points, $A, \ldots, G$ be related as suggested in the accompanying model. The following are true statements.

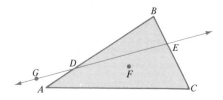

a. $\overleftrightarrow{DE} \cap \triangle ABC = \{D,E\}$.   b. $\overleftrightarrow{DE} \cap \text{Reg}(\triangle ABC) = \overline{DE}$.
c. $F \in \text{Int}(\triangle ABC)$.   d. $F \in \blacktriangle ABC$.
e. $B \in \blacktriangle ABC$.   f. $B \notin \text{Int}(\triangle ABC)$.
g. $\triangle ABC$ is the boundary of $\blacktriangle ABC$.

The accompanying model is of $\blacktriangle ABC \cap \overleftrightarrow{DE}$:$B$. (The dashed model line segment DE is to suggest that the points of $\overline{DE}$ are not included in the region.)

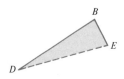

Exercise set 7.1

1. Let the geometric figures be related as indicated in the accompanying model.
    A. Where appropriate, for each of the following statements, indicate whether it is true or false; in other cases complete the sentence to make it a true statement.

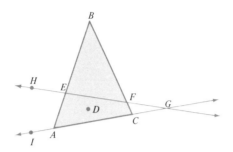

a. $D \in \text{Int}(\triangle ABC)$.

b. $\overline{AB} \subseteq \blacktriangle ABC$.

c. $E \in \blacktriangle ABC$.

d. $\overleftrightarrow{AG} \cap \triangle ABC = ?$.

e. $\overleftrightarrow{AG} \cap \blacktriangle ABC = ?$.

f. $\overleftrightarrow{EF} \cap \triangle ABC = ?$.

g. $\overleftrightarrow{EF} \cap \text{Int}(\triangle ABC) = ?$.

h. $\overleftrightarrow{EF} \cap \text{Ext}(\triangle ABC) = \overrightarrow{EH} \cup \overrightarrow{FG}$.

i. $\overleftrightarrow{EF} \cap \text{Ext}(\triangle ABC) = \overleftrightarrow{EF} \setminus \overline{EF}$.

j. $\overleftrightarrow{AG} \cap \text{Int}(\triangle ABC) = ?$.

k. $\overleftrightarrow{AG} \cap \text{Ext}(\triangle ABC) = ?$.

l. $\overline{EF} \cap \blacktriangle ABC = ?$.

m. $\overleftrightarrow{BC} \cap \blacktriangle ABC = \varnothing$.

n. $E \in \triangle ABC$.

o. $D \in \text{Ext}(\triangle ABC)$.

p. $D \in \blacktriangle ABC$.

q. $\overline{DF} \subseteq \blacktriangle ABC$.

B. Draw models of each of the following:

a. $\blacktriangle ABC \cap \overleftrightarrow{EF}{:}B$ (that is, the intersection of the triangular region ABC and the B-side of $\overleftrightarrow{EF}$).

b. $\blacktriangle ABC \cap \overleftrightarrow{EF}{:}A$.   c. $\blacktriangle ABC \cap \overleftrightarrow{AC}{:}B$.

d. $\overleftrightarrow{BC}{:}G \cap \triangle CFG$.

2. Draw models to illustrate the following. (All distinct letters name distinct points.)

a. Interior of $\angle ABC$.

b. $\overline{EF} \cap \angle ABC = \{G,H\}$.

c. $\overline{EF} \cap \text{Int}(\angle ABC) \subset \overline{EF}$.

d. $\overleftrightarrow{EF} \cap \angle ABC = \{B\}$.

e. $\overline{EF} \cap \angle ABC = \{E\}$.

f. $\overleftrightarrow{EF} \cap \angle ABC = \overline{EB}$.

g. $\overline{EF} \cap \text{Int}(\angle ABC) = \overline{EF}$.

h. $\overline{EF} \cap \text{Ext}(\angle ABC) = \overline{EF}$.

i. $\overline{EF} \cap \text{Int}(\angle ABC) = \overline{EF} \setminus \{E\}$.

j. $\overrightarrow{DG} \cap \text{Int}(\angle ABC) = \overrightarrow{DG}$.

k. $\overrightarrow{DG} \cap \text{Int}(\angle ABC) = \overset{\circ}{\overrightarrow{DG}}$.

l. $\triangle ABC \cap \overleftrightarrow{EF} = \{E,F\}$.

m. $\triangle ABC \cap \overleftrightarrow{EF} = \overline{AB}$. **n.** $\triangle ABC \cap \overleftrightarrow{EF} = \{C,P\}$.

o. $\text{Ext}(\triangle ABC) \cap \overrightarrow{PR} = \overset{o}{\overrightarrow{PR}}$. **p.** $\text{Ext}(\triangle ABC) \cap \text{Int}(\angle ABC)$.

q. $\text{Int}(\angle ABC) \cap \overleftrightarrow{AC}{:}B$. **r.** $\blacktriangle ABC \setminus \triangle ABC$.

s. $\blacktriangle ABC \cap \overline{EF} \subset \overline{EF}$. **t.** $\overline{EF} \cap \blacktriangle ABC \subset \text{Int}(\triangle ABC)$.

u. $\blacksquare ABCD \cap \overleftrightarrow{AC}{:}D$, where $\overline{AC}$ is a diagonal of $\square ABCD$.

v. $(\blacksquare ABCD \cap \overleftrightarrow{AC}{:}B) \cup \overline{AC}$, where $\overline{AC}$ is a diagonal of $\square ABCD$.

w. $\overleftrightarrow{EF} \cap \triangle ABC$ is a set containing one point.

x. $\overrightarrow{EF} \cap \triangle ABC$ is a set containing one point.

y. $\overline{EF} \cap \triangle ABC$ is an infinite set and $\overline{EF} \nsubseteq \triangle ABC$.

7.2 MEASURE OF REGIONS

We will discuss some notions related to the measure of models of regions; this should give some motivation for the precise notion of measure of geometric regions. In discussing the ideas related to approximating the measure of models of line segments and angles, we first needed to have some unit as a basis for comparison. For line segments, a unit line segment serves as the basis for comparison; for angles, a unit angle serves this purpose. It seems natural, therefore, that we must choose a unit region as the basis for comparison to approximate the measure of model regions. We choose as our unit of comparison a model region which we consider to be simpler than others; namely, a unit-square region. A unit-square region is a region whose boundary is a square and each side of the square is a unit line segment.

Through a sequence of models we will illustrate the procedure by which we can approximate the measure of a model of a rectangular region. We let the accompanying figure be our model of the unit-square region. The measure of the unit-square region

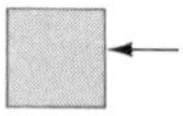

◄— model of unit-square region

is $^+1$. If we place four models of unit-square regions side to side, as in the accompanying figure, it is clear that the measure of the

region which is the union of these four regions is ⁺4.   Now if we place three such rows of model square regions, as in the accompanying figure, the measure of the model rectangular region is ⁺12 (that is, ⁺3 × ⁺4).

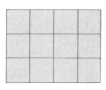

In the above illustration we chose a model unit-square region and built a model rectangular region using more models of the unit-square region.   In this way we were able to determine quite accurately the measure of the resulting rectangular region.   However, as illustrated in Examples 7.3 and 7.4, it is not always possible to be quite this accurate.

EXAMPLE 7.3 We use the accompanying model unit-square

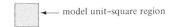 ◄—— model unit–square region

region to approximate the measure of the accompanying model of

a rectangular region.   We proceed as suggested in the accompanying model.   We conclude that the measure of the model of the

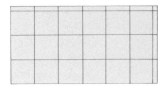

rectangular region is greater than $^+18$ and less than $^+28$. That is, the square unit (or unit-square) measure of the region is between $^+18$ and $^+28$.

EXAMPLE 7.4 We use the model unit-square region to approxi-

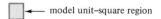 — model unit–square region

mate the measure of the accompanying model region. We proceed

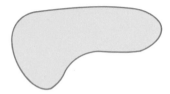

as illustrated in the accompanying model. We can conclude that

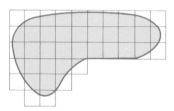

the measure of the model region, with respect to the unit-square region, is greater than $^+17$ and less than $^+41$.

7.2.1 Measure of rectangular regions

The examples in the previous section gave motivation for some of the necessary considerations for determining the exact measure of geometric figures (sets of points).

Postulate 7.1 (Region measure postulate) There exists a measure function m which assigns to every region a unique positive real number relative to a unit-square region, which is assigned $^+1$. If $\mathscr{R}$ denotes any region, then the real number $m(\mathscr{R})$ is the *measure* of $\mathscr{R}$.

In Example 7.4, we were not able to get a very accurate approximation of the measure of the model region. If we were to choose smaller and smaller unit-square regions, we could obtain better and better approximations. Such considerations concerning line segments and angles led us to relate the measure of line segments and angles to line and ray coordinate systems, respectively. We give Example 7.5 to suggest that the area of a rectangular region can be related to a rectangular coordinate system.

EXAMPLE 7.5 We consider the accompanying model of a rec-

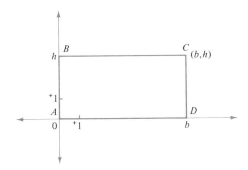

tangle *ABCD*. Since a rectangle is a union of line segments and contains four right angles, we can choose line coordinate systems on the lines which contain a pair of adjacent sides to form a rectangular coordinate system with (0,0) assigned to the vertex. Then, $m(\overline{AD}) = |b - 0| = b$ and $m(\overline{AB}) = |h - 0| = h$ relative to our chosen unit-pair. The measure of ■*ABCD* relative to the unit-square region is $b \times h$.

Definition 7.2 For every rectangle *ABCD*, the measure of the rectangular region *ABCD* is given by

$$m(\blacksquare ABCD) = bh,$$

where b and h are the measures of a pair of adjacent sides of $\square ABCD$.

We will use the definition of the measure of a rectangular region to derive the formulas to determine the measure for other types of polygonal regions; particularly, triangular and quadrilateral regions. In these derivations we will use the following two postulates.

Postulate 7.2 (Region measure addition postulate) If two co-planar regions $\mathscr{R}_1$ and $\mathscr{R}_2$, determined by simple closed curves $\mathscr{C}_1$ and $\mathscr{C}_2$, are such that $\text{Int}(\mathscr{C}_1) \cap \text{Int}(\mathscr{C}_2) = \varnothing$, then $m(\mathscr{R}_1 \cup \mathscr{R}_2) = m(\mathscr{R}_1) + m(\mathscr{R}_2)$.

Postulate 7.3 For all simple closed curves $\mathscr{C}_1$ and $\mathscr{C}_2$, if $\mathscr{C}_1 \cong \mathscr{C}_2$, then $m(\text{Reg } \mathscr{C}_1) = m(\text{Reg } \mathscr{C}_2)$.

An important result which follows from Postulates 7.2 and 7.3 is illustrated in Example 7.6.

EXAMPLE 7.6 Consider the accompanying model of ▬*ABCD*.

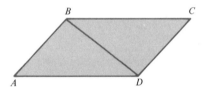

Points *B* and *D* determine a line segment; and therefore, to-gether with the sides of ⬜*ABCD*, determine the two triangles *ABD* and *CDB*. Also the line segment *BD* forms the two triangu-lar regions *ABD* and *CDB*. $\text{Int}(\triangle ABD) \cap \text{Int}(\triangle CDB) = \varnothing$; hence, $m(\text{▬}ABCD) = m(\blacktriangle ABD) + m(\blacktriangle CDB)$. Moreover, since the diag-onal $\overline{BD}$ forms with the remaining sides the two congruent tri-angles, it follows that $m(\blacktriangle ABD) = m(\blacktriangle CDB)$.

A remark The word *area* is used in reference to regions and the measure of regions as the words *length* and *perimeter* are used in reference to line segments and polygons. That is, a measure of a region is a real number determined by comparing the given re-gion to a given unit; the *area* of a region is a measure of the region together with the unit of measure. We remark on this concept again in Section 7.3.

Exercise set 7.2

1. Using the accompanying model unit-square region, approxi-

unit–square region

mate the measure of the following model regions.

a.

b.

c.

d.

2. Using the accompanying model unit-square region, approximate the measure of the model regions in Exercise 1.

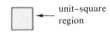
unit–square region

3. Determine the measure of the rectangular region $ABCD$ when:

a. $m(\overline{AB}) = {}^+4$ and $m(\overline{BC}) = {}^+3$.

b. $m(\overline{AD}) = \frac{{}^+2}{{}^+3}$ and $m(\overline{DC}) = \frac{{}^+3}{{}^+2}$.

c. $m(\overline{BC}) = {}^+9$ and $m(\overline{CD}) = {}^+3$.

7.3 THE PRACTICE OF APPROXIMATING THE MEASURE OF REGIONS

We will consider specifically the g.p.e. in approximating the measure of a model of a rectangular region using the formula bh, where b and h are the approximate measures of two adjacent sides of the model rectangle. This consideration will generalize to the consideration of the g.p.e. in other situations in which an approximation of a measure is obtained by multiplying the approximation of other measures.

Consider the accompanying model of ■$ABCD$.   Suppose we

determine that $m(\overline{AB}) \doteq {}^+2.3$ and $m(\overline{AD}) \doteq {}^+4.4.$ Then,

${}^+2.25 \leqslant m(\overline{AB}) < {}^+2.35$ (g.p.e. is ${}^+0.05$),

${}^+4.35 \leqslant m(\overline{AD}) < {}^+4.45$ (g.p.e. is ${}^+0.05$).

From the fact that $m(■ABCD) = m(\overline{AB}) \times m(\overline{AD})$ and from properties of inequalities, we have

${}^+2.25 \times {}^+4.35 \leqslant m(■ABCD) < {}^+2.35 \times {}^+4.45$

or

${}^+9.7875 \leqslant m(■ABCD) < {}^+10.4575.$

Using the formula bh we determine that

$m(■ABCD) \doteq {}^+2.3 \times {}^+4.4;$

that is,

$m(■ABCD) \doteq {}^+10.12.$

Now, ${}^+10.12 - {}^+9.7875 = {}^+0.3325$ and ${}^+10.4575 - {}^+10.12 = {}^+0.3375$; thus, the possible error is greater "in one direction than in the other."   However, it is easily observed that the g.p.e. of the approximation of $m(■ABCD)$ (the maximum of ${}^+0.3325$ and ${}^+0.3375$) obtained by multiplying the approximate measures of the sides of the model of □$ABCD$ *is greater* than the g.p.e. of each of these approximations (${}^+0.05$).   We leave to the reader the problem of computing

$$\frac{{}^+0.3375}{{}^+10.12} - \left(\frac{{}^+0.05}{{}^+2.3} + \frac{{}^+0.05}{{}^+4.4}\right)$$

to observe that the r.e. of the approximation of $m(■ABCD)$ is very near the sum of the r.e.'s of the two factors.   We prove these results in general as follows: Let

$m(\overline{AB}) \doteq x$ with g.p.e. $= a$,
$m(\overline{BC}) \doteq y$ with g.p.e. $= b$.

Then,

$x - a \leqslant m(\overline{AB}) < x + a$,
$y - b \leqslant m(\overline{BC}) < y + b$.

Now, using properties of inequalities and multiplication, we have

$$(x - a)(y - b) \leqslant m(\overline{AB}) \times m(\overline{BC}) < (x + a)(y + b)$$

so that

$$xy - xb - ya + ab \leqslant m(\overline{AB}) \times m(\overline{BC}) < xy + ay + bx + ab.$$

From this we see that $m(\overline{AB}) \times m(\overline{BC})$ can differ from xy by at most $(xy + ay + bx + ab) - xy$ or $ay + bx + ab$. Now since a and b are small compared to x and y, the product ab will be very small compared to x and y (in our illustration above $ab = {}^+0.0025$); thus in practice, we can ignore ab and consider $ay + bx$ to be the g.p.e. of the approximation of $m(\overline{AB}) \times m(\overline{BC})$. (In our illustration, ignoring ab changes the number we accept as the g.p.e. from ${}^+0.3375$ to ${}^+0.3350$.) Therefore, we accept that

$$m(\overline{AB}) \times m(\overline{BC}) = m(\blacksquare ABCD) \doteq xy$$

with

g.p.e. $= ay + bx$.

Then, the r.e. of this approximation is

$$\frac{ay + bx}{xy} = \frac{ay}{xy} + \frac{bx}{xy} = \frac{a}{x} + \frac{b}{y}.$$

That is, the relative error of an approximation determined by taking the product of approximations is the sum of the relative errors of the factors.

The above illustration and proof are not intended to suggest that whenever we approximate a measure by multiplying approximations we must calculate the g.p.e. and r.e. of all approximations involved. The reader, however, must be aware that the approximations obtained in this manner are less precise (and less accurate) than the approximate measures which are the factors. The reader will recall that the same is true for approximations obtained by adding approximations of measures.

In order not to suggest more precision (and accuracy) than is warranted from the approximations involved in a computation, a rule of thumb is generally adopted which is stated in terms of significant digits (those digits which have real meaning in the approximation).

Rule   When determining an approximation of a measure by adding or multiplying approximate measures, we should "round off" the sum or product to the same number of *significant digits* as the addend or factor which has the least number of significant digits.

This rule does not give the exact g.p.e. (or r.e.) of a "computed" approximation; in fact it usually errs on the conservative side; that is, this rule will never suggest greater precision or accuracy than is possible from the approximations involved in the computation. We illustrate an application of this idea by considering the accom-

panying model of ■*ABCD*.   Suppose that with respect to an inch unit

$m(\overline{AB}) \doteq {}^{+}3.9,$

$m(\overline{AD}) \doteq {}^{+}6.5.$

Then, $m(\square ABCD) \doteq {}^{+}3.9 + {}^{+}3.9 + {}^{+}6.5 + {}^{+}6.5$; that is,

$m(\square ABCD) \doteq {}^{+}20.8.$

Since each of the addends is named by a numeral with two significant digits we round off the approximation of $m(\square ABCD)$ to two significant digits and report that $m(\square ABCD) \doteq {}^{+}21$ and that the perimeter of $\square ABCD$ is approximately $^{+}21$ inches.

To approximate the measure of the rectangular region using the formula bh we have

$m(\blacksquare ABCD) \doteq {}^{+}3.9 \times {}^{+}6.5;$

therefore,

$m(\blacksquare ABCD) \doteq {}^+25.35.$

Since each of the factors is named by a numeral with two significant digits we round off the approximation of $m(\blacksquare ABCD)$ to two
significant digits and report that $m(\blacksquare ABCD) \doteq {}^+25$ and that the
area is approximately ${}^+25$ *inch-squares,* or what is more customary
we report that the *area* is approximately ${}^+25$ square inches. We
again call attention to the distinction between a measure of a region
and the area of a region; a *measure* is a real number determined by
some unit of measure, the *area* is a measure together with the unit
used to determine that real number.

Exercise set 7.3

1. The questions below refer to the accompanying model of a
    rectangular region.

a. Let $m(\overline{AB}) \doteq {}^+3.4$ and $m(\overline{AD}) \doteq {}^+5.6.$

    (1) _____ $\leqslant m(\overline{AB}) <$ _____.

    (2) _____ $\leqslant m(\overline{AD}) <$ _____.

    (3) $m(\square ABCD) \doteq$ _____.

    (4) _____ $\leqslant m(\square ABCD) <$ _____.

    (5) Determine the g.p.e. and r.e. of the approximations
        in (1), (2), and (4).

    (6) Express $m(\square ABCD) \doteq$ _____ with the appropriate
        number of significant digits.

    (7) $m(\blacksquare ABCD) \doteq$ _____.

    (8) _____ $\leqslant m(\blacksquare ABCD) <$ _____.

    (9) Compare the g.p.e. and r.e. of the approximations in
        (1), (2), and (8).

    (10) Express $m(\blacksquare ABCD) \doteq$ _____ with the appropriate
         number of significant digits.

b. Repeat (1)–(10) above for $m(\overline{AB}) \doteq {}^+3.40$ and $m(\overline{AD}) \doteq {}^+5.61$.

2. Let $\mathscr{C}$ be a circle whose radius is approximately ${}^+2.5$. Use ${}^+3.14$ as an approximation for π. Approximate the measure of the circle and determine the g.p.e. and r.e. of this approximation.

7.4 MEASURE OF TRIANGULAR REGIONS

In order to discuss the measure (and area) of triangular regions we introduce, in Example 7.7, some additional vocabulary related to triangles.

EXAMPLE 7.7 Consider the triangles suggested by the following models.

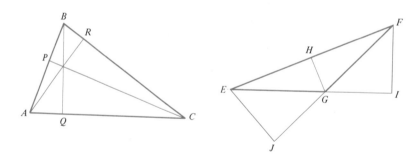

a. $\triangle ABC$: If $\overline{AR} \perp \overleftrightarrow{BC}$, $\overline{PC} \perp \overleftrightarrow{AB}$, and $\overline{BQ} \perp \overleftrightarrow{AC}$, then:
(1) $\overline{AR}$ is the *altitude* of $\triangle ABC$ to the line BC.
(2) $\overline{BQ}$ is the *altitude* of $\triangle ABC$ to the line AC.
(3) $\overline{PC}$ is the *altitude* of $\triangle ABC$ to the line AB.

b. $\triangle EFG$: If $\overleftrightarrow{GH} \perp \overleftrightarrow{EF}$, $\overline{FI} \perp \overleftrightarrow{EG}$, $\overline{EJ} \perp \overleftrightarrow{GF}$, then:
(1) $\overline{GH}$ is the *altitude* of $\triangle EFG$ to the line EF, $\overline{EF}$ is a *base* of $\triangle EFG$, and $\overline{GH}$ the corresponding altitude.
(2) $\overline{FI}$ is the *altitude* of $\triangle EFG$ to the line EG, $\overline{EG}$ is a *base* of $\triangle EFG$, and $\overline{FI}$ the corresponding altitude.
(3) $\overline{EJ}$ is the *altitude* of $\triangle EFG$ to the line GF, $\overline{GF}$ is a *base* of $\triangle EFG$, and $\overline{EJ}$ the corresponding altitude.

Example 7.7 illustrates Definition 7.3, given on the next page.

Definition 7.3 An *altitude* of a triangle is the line segment from the vertex that is perpendicular to the line which contains the side opposite the vertex. The side contained by this line is a *base* of the triangle.

We will derive the measure for a triangular region in a sequence starting with a right triangular region. We consider the rectangular region *ABCD* shown in the accompanying model. Let

$b = m(\overline{AD})$ and $h = m(\overline{AB})$. The diagonal $\overline{BD}$ shown in the accompanying model forms with the sides of $\square ABCD$ two congruent

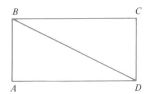

right triangles. That is, $\triangle ABD \cong \triangle CDB$; therefore, by Postulate 7.3,

$m(\blacktriangle ABD) = m(\blacktriangle CDB).$

Also, by the region measure addition postulate it follows that

$m(\blacksquare ABCD) = m(\blacktriangle ABD) + m(\blacktriangle CDB)$

so that we have

$m(\blacksquare ABCD) = m(\blacktriangle ABD) + m(\blacktriangle CDB) = {}^{+}2 \times m(\blacktriangle ABD).$

$\therefore m(\blacktriangle ABD) = \dfrac{{}^{+}1}{{}^{+}2} \times m(\blacksquare ABCD).$

$\therefore m(\blacktriangle ABD) = \dfrac{{}^{+}1}{{}^{+}2} bh.$

That is, the measure of a right triangular region is one-half the product of the measure of a base and the measure of the altitude to that base.

Next we consider a triangle as in the accompanying model such

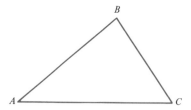

that each of its angles is acute.　The altitude from any vertex, as in the accompanying model, forms with the sides of the triangle two right triangles.[1]

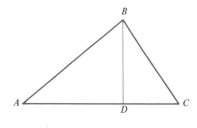

$$m(\blacktriangle ADB) = \frac{+1}{+2} \times m(\overline{AD}) \times m(\overline{DB}).$$

$$m(\blacktriangle DBC) = \frac{+1}{+2} \times m(\overline{DC}) \times m(\overline{DB}).$$

$$m(\overline{AD}) + m(\overline{DC}) = m(\overline{AC}).[2]$$

$$\begin{aligned} m(\blacktriangle ABC) &= m(\blacktriangle ADB) + m(\blacktriangle DBC) \\ &= \frac{+1}{+2} \times (m(\overline{AD}) + m(\overline{DC})) \times m(\overline{BD}) \\ &= \frac{+1}{+2} \times m(\overline{AC}) \times m(\overline{BD}). \end{aligned}$$

[1] This fact can be proved in a more detailed and complete development of geometry.

[2] This assumes A–D–C; this fact would be established if we were to undertake the proof of footnote 1.

That is, the measure of the triangular region is $\frac{+1}{+2}\, bh$, where b is the measure of a base and h the measure of the altitude to that base.

Finally, if $\triangle ABC$ has an obtuse angle, we leave to the reader (Exercise 1 of Exercise set 7.4) to show that $m(\blacktriangle ABC) = \frac{+1}{+2}\, bh$, where b is the measure of a base and h is the measure of the altitude to that base. Thus, we have outlined a proof of Theorem 7.1.

Theorem 7.1 For every triangle ABC such that the measure of one base is b and the measure of the altitude to the line containing that base is h, then the measure of the triangular region determined by $\triangle ABC$ is given by $m(\blacktriangle ABC) = \frac{+1}{+2}\, bh$.

7.5 MEASURE OF OTHER REGIONS

Now that we have a method to determine the measure of every triangular region, we are in a position to develop formulas for determining the measure of other polygonal regions. We will restrict ourselves to particular quadrilaterals. Some exercises in Exercise set 7.4 suggest procedures for other polygons.

Definition 7.4 An *altitude* of a parallelogram is a line segment with at least one endpoint on a side of the parallelogram, is perpendicular to the line containing that side, and the other endpoint is in the line containing the opposite side. Each side of the parallelogram which contains an endpoint of an altitude is a *base* of the parallelogram.

Definition 7.4 is illustrated in Example 7.8.

EXAMPLE 7.8 Using the accompanying models, we see that:

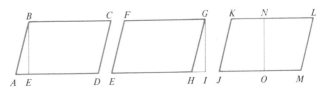

If $\overline{BE} \perp \overleftrightarrow{BC}$, then $\overline{BE}$ is an altitude of $\square ABCD$ and $\overline{BC}$ and $\overline{AD}$ are bases.

If $\overline{GI} \perp \overleftrightarrow{FG}$, then $\overline{GI}$ is an altitude of $\square EFGH$ and $\overline{FG}$ is a base.

If $\overline{NO} \perp \overleftrightarrow{KL}$, then $\overline{NO}$ is an altitude of $\square JKLM$ and $\overline{KL}$ and $\overline{JM}$ are bases.

Definition 7.5   An *altitude* of a trapezoid is a line segment which has one endpoint on one of the opposite parallel sides, is perpendicular to the line which contains this side, and its other endpoint is in the line containing the opposite side.   The sides contained by these lines are *bases* of the trapezoid.

Theorem 7.2   For every trapezoid $ABCD$, the measure of the trapezoidal region $ABCD$ is given by

$$m(\blacksquare ABCD) = \frac{+1}{+2} (b_1 + b_2)h$$

where b_1 and b_2 are the measures of the two bases and h is the measure of the altitudes to these bases.

The proof of this is left as an exercise for the reader.

Finally we turn our attention to the measure of circular regions. The reader may recall that for a circle with radius r the measure of the circular region it determines is given by πr^2.   Since it requires techniques of calculus to *prove* this, we will limit our discussion to some diagrams which suggest the plausibility of the formula.

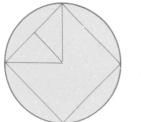  

FIG. 7.1
Models to suggest the measure of a circular region.

Consider the models in Fig. 7.1.   Let the radius of each circle be r.   For each triangle determined by the center and consecutive vertices of the "inscribed" polygons, let h be the measure of its altitude and b the measure of its base.   Then the measures of the inscribed polygonal regions are, respectively,

$^+4 \cdot \dfrac{^+1}{^+2} \, bh,$ $^+8 \cdot \dfrac{^+1}{^+2} \, bh,$ and $^+16 \cdot \dfrac{^+1}{^+2} \, bh,$

or

$^+4b \cdot \dfrac{^+1}{^+2} \, h,$ $^+8b \cdot \dfrac{^+1}{^+2} \, h,$ and $^+16b \cdot \dfrac{^+1}{^+2} \, h.$

The measures p of the inscribed polygons are, respectively, ^+4b, ^+8b, and ^+16b. Therefore, the measures of the inscribed polygonal regions are all equal to $\dfrac{^+1}{^+2} \, ph$. Now, as the number of sides of the polygon increases the measure of the polygon approaches the measure—$^+2\pi r$—of the circle, and h approaches r. Therefore, the measure, $\dfrac{^+1}{^+2} \, ph$, of the inscribed polygonal region approaches

$$\dfrac{^+1}{^+2} \cdot {^+2\pi r} \cdot r = \pi r^2.$$

Exercise set 7.4

1. Suppose $\triangle ABC$ has an obtuse angle. With the aid of the accompanying model, determine the formula for the measure of $\triangle ABC$.

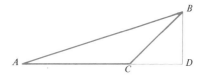

2. In $\triangle DEF$, $m(\overline{DF}) = {^+6}$, $\overline{EG}$ is an altitude, and $m(\overline{EG}) = {^+4}$. In $\triangle HIJ$, $m(\overline{HJ}) = {^+8}$ and $\overline{IK}$ is an altitude. If $m(\blacktriangle DEF) = m(\blacktriangle HIJ)$, determine $m(\overline{IK})$. (See the accompanying models.)

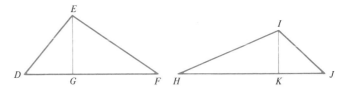

3. Determine the measure of a square region such that the measure of each of its sides is $^+3$.

4. In △*ABC*, $\overline{CD}$ is the altitude to $\overleftrightarrow{AB}$ and $\overline{AE}$ is the altitude to $\overleftrightarrow{BC}$, as in the accompanying model.

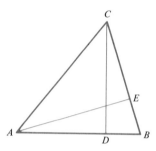

a. If $m(\overline{AB}) = {}^+8$, $m(\overline{CD}) = {}^+9$, $m(\overline{AE}) = {}^+12$, determine $m(\overline{BC})$.

b. If $m(\overline{BC}) = {}^+7$, $m(\overline{AE}) = {}^+9$, $m(\overline{AB}) = {}^+10$, determine $m(\overline{CD})$.

5. In △*ABC* and □*DEFG*, as in the accompanying models, $m(\overline{AB}) = m(\overline{DG})$ and $m(\blacktriangle ABC) = m(\blacksquare DEFG)$. If $\overline{CG}$ and $\overline{EH}$ are altitudes, how does $m(\overline{CG})$ compare with $m(\overline{EH})$?

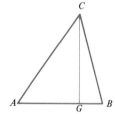
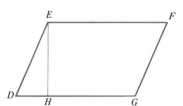

6. Consider circles $\mathscr{C}_1$ and $\mathscr{C}_2$ with radii ${}^+6$ and ${}^+4$, respectively. Determine the measure of the shaded region in the following model.

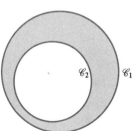

7. Consider □$ABCD$ with parallel sides $\overline{AB}$ and $\overline{CD}$, and altitudes $\overline{DE}$ and $\overline{CF}$, as shown in the accompanying model.

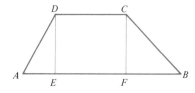

a. If $m(\overline{DE}) = {}^+5$, $m(\overline{DC}) = {}^+6$, and $m(\overline{AB}) = {}^+12$, determine $m(\blacksquare ABCD)$.

b. If $m(\overline{AE}) = {}^+3$, $m(\overline{AB}) = {}^+12$, $m(\overline{DE}) = {}^+4$, and $m(\overline{FB}) = {}^+4$, determine:

    (1) $m(\blacksquare DCFE)$ (2) $m(\blacktriangle BFC)$ (3) $m(\blacktriangle ADE)$

c. If □$DCFE$ is a square, $m(\blacksquare ABCD) = {}^+40$, $m(\blacktriangle AED) = {}^+9$, and $m(\blacktriangle BCF) = {}^+15$, determine:

    (1) $m(\blacksquare DCFE)$ (2) $m(\overline{DE})$

8. If □$ABCD$ is a rhombus, $\overline{BE}$ is an altitude, $m(\overline{AD}) = {}^+5$, $m(\overline{BE}) = {}^+4$, and $m(\overline{AE}) = {}^+3$, as in the accompanying model, determine:

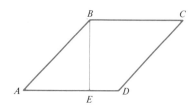

a. $m(\blacksquare ABCD)$

b. the measure of the quadrilateral region determined by $\overline{BC} \cup \overline{CD} \cup \overline{DE} \cup \overline{EB}$

c. $m(\overline{DC})$

9. In the model (next page): If $m(\overline{AB}) = {}^+5$, $m(\overline{BC}) = {}^+5$, $m(\overline{DF}) = {}^+4$, $m(\overline{BG}) = {}^+4$, $m(\overline{AG}) = {}^+3$, $m(\overline{GC}) = {}^+3$, $m(\overline{BE}) = {}^+6$, $\overline{BG} \perp \overline{AC}$, and $\overline{BE} \perp \overline{FD}$, determine:

a. $m(\blacktriangle ABC)$ b. $m(\blacktriangle ABG)$

c. $m(\blacktriangle BGC)$ d. $m(\blacksquare ACDF)$

e. the area of the region bounded by $\overline{AB} \cup \overline{BC} \cup \overline{CD} \cup \overline{DF} \cup \overline{FA}$

7   regions, measure of regions

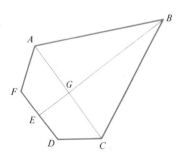

10. Suppose that $m(\blacktriangle ABC) = m(\blacktriangle DEF)$.   Is it necessarily true
     that $\triangle ABC \cong \triangle DEF$?   Prove your answer.

11. In $\square ABCD$, as in the accompanying model, $\overline{DE}$ is an altitude,
     $\overline{CG}$ is an altitude, $m(\blacktriangle AED) = m(\blacktriangle BGC)$, $m(\overline{AD}) = {}^+6$,
     $m(\overline{ED}) = {}^+4$, and the measure of $\square ABCD$ is ${}^+32$.   Deter-
     mine $m(\blacksquare ABCD)$.

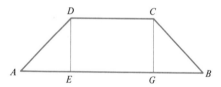

12. Prove: The measure of a region determined by a rhombus is
     equal to one-half the product of the measures of the two di-
     agonals of the rhombus.

subsets of space— surfaces and solids

8.1 DEFINITION OF SURFACE AND SPHERE

A plane is related to space as a line is related to the plane which contains it. This relationship is given specifically in Postulate 8.1.

Postulate 8.1 (Space separation postulate) For every plane α, α separates the points in space not on α into two convex sets such that if P is in one of the sets and Q is in the other, then $\overline{PQ} \cap \alpha \neq \varnothing$.

The model in Fig. 8.1 suggests the conditions of the postulate.

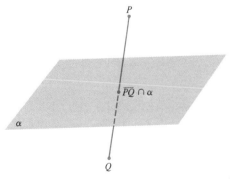

FIG. 8.1
A plane separates space.

A surface is to space as a curve is to a plane. The unit-line segment is the prototype of curves; the space analog is the unit square. Therefore, it would seem natural to define a surface as in Definition 8.1.

Definition 8.1 A *surface* is the image of a continuous function from the unit square to space.

A similar definition for curve led to definitions of simple curves and simple closed curves. Surfaces can be similarly classified using Definition 8.1. However, we consider such an approach beyond the scope of this text. Therefore, we will limit our discussion to surfaces which are relatively easy to define.

Definition 8.2 A *sphere* is a set $\mathscr{F}$ consisting of all points X in space for which there exists a point P and a real number r, $r > 0$, such that for every point $X \in \mathscr{F}$, $PX = r$. The point P is called the *center* of the sphere and r is called the *radius*.

Definition 8.3 The *interior* of a sphere with radius r is the set of all points X in space whose distance from the center is less than r.

Definition 8.4 A subset $\mathscr{A}$ of space is *bounded* if and only if there exists a sphere $\mathscr{F}$ such that $\mathscr{A} \subseteq \text{Int}(\mathscr{F})$.

Exercise

Use Definition 8.4 to determine whether:

a. a line segment is a bounded set
b. a triangle is a bounded set
c. a circle is a bounded set
d. a line is a bounded set

8.2 DEFINITION AND CLASSIFICATION OF POLYHEDRONS

We next direct our attention to the study of simple closed surfaces which are the union of polygonal regions.

Definition 8.5 A *polyhedron* is the union of a finite number of polygonal regions such that:

1. The intersection of every pair of the polygonal regions contains no points interior to the boundaries of the regions.

2. Every side of each polygonal region is the side of exactly two polygonal regions.

The polygonal regions in the union are called the *faces* of the polyhedron. The sides of the regions are called the *edges*, and the vertices of the regions are called *vertices* of the polyhedron. Some models of polyhedrons are shown in Fig. 8.2.

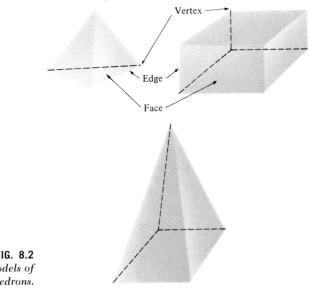

Vertex

Edge

Face

FIG. 8.2
Models of
polyhedrons.

It is possible to prove that every polyhedron $\mathscr{F}$ separates the set of points in space not on the polyhedron into two disjoint subsets $\mathscr{A}$ and $\mathscr{B}$ such that $\mathscr{A}$ is bounded and $\mathscr{B}$ is unbounded. (Since the proof will not add substantially to the reader's understanding of the result, we omit it here.) The model in Fig. 8.3

$\mathscr{B}$

FIG. 8.3
A polyhedron
separates the re-
mainder of space
into two sets, one
bounded and one
unbounded.

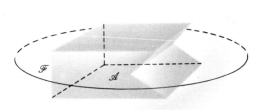

$\mathscr{F}$

$\mathscr{A}$

may help the reader to visualize this result. It suggests that $\mathscr{A}$ is bounded because there exists a sphere which contains $\mathscr{A}$.

Definition 8.6 For every polyhedron $\mathscr{F}$, the *interior* of $\mathscr{F}$, denoted by Int($\mathscr{F}$), is the bounded set $\mathscr{A}$ described above.

Definition 8.7 For every polyhedron $\mathscr{F}$, the *exterior* of $\mathscr{F}$, denoted by Ext($\mathscr{F}$), is the unbounded set $\mathscr{B}$ described above.

A polyhedron is called *convex* if and only if its interior is a convex set. Models to illustrate polyhedrons which are and are not convex are given in Fig. 8.4.

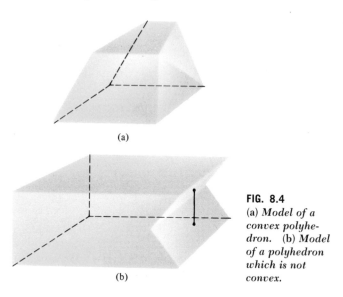

(a)

(b)

FIG. 8.4
(a) *Model of a convex polyhedron.* (b) *Model of a polyhedron which is not convex.*

A characterization of the interior and exterior for some polyhedrons which is based on whether certain rays "cross" the polyhedron an odd or even number of times is possible, and is quite similar to the characterization given for the interior and exterior of some simple closed curves. Such a characterization requires a consideration of what is meant by a ray "crossing" a region. We give the hint that spheres will be involved in this discussion in a manner analogous to the way circles are involved in the discussion of rays crossing curves and ask the reader to accept the challenge of thinking this through. We give the following even simpler characterization for convex polygons.

For every convex polyhedron $\mathscr{F}$, the *interior* of $\mathscr{F}$ is the set of all points $P \in S$, $P \notin \mathscr{F}$, such that for every $X \in S$, $\overrightarrow{PX} \cap \mathscr{F} \neq \varnothing$. For every convex polyhedron $\mathscr{F}$, the *exterior* of $\mathscr{F}$ is the set of all points $Q \in S$, $Q \notin \mathscr{F}$, such that there exists a point $X \in S$ for which $\overrightarrow{QX} \cap \mathscr{F} = \varnothing$.

The model in Example 8.1 illustrates this characterization for a convex polyhedron.

EXAMPLE 8.1 Consider the accompanying model of the polyhedron $\mathscr{F}$.

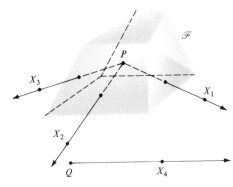

$$\overrightarrow{PX_1} \cap \mathscr{F} \neq \varnothing, \qquad \overrightarrow{PX_2} \cap \mathscr{F} \neq \varnothing, \qquad \overrightarrow{PX_3} \cap \mathscr{F} \neq \varnothing, \ldots ;$$

this *suggests* that $P \in \text{Int}(\mathscr{F})$. Since it is apparent that $\overrightarrow{QX_4} \cap \mathscr{F} = \varnothing$, $Q \in \text{Ext}(\mathscr{F})$.

Classification of polyhedrons can be accomplished in a manner which is analogous to the way polygons are classified; that is, according to certain relationships which exist between the faces of the polyhedron. The reader will recall that the faces of a polyhedron are *regions bounded by polygons*. A polyhedron is a *regular polyhedron* if and only if the boundaries of all of its faces are congruent regular polygons.

Definition 8.8 A polyhedron is a *prism* if and only if

1. Two of its faces are bounded by congruent polygons which are in parallel planes.

2. The remaining faces of the polyhedron are regions bounded by parallelograms.

   Any two parallel faces of a prism are called *bases;* the remaining faces are called *lateral faces.*   The intersection of two lateral faces is called a *lateral edge.*   The union of the lateral faces is called the *lateral surface.*

   Further classification of prisms is as follows:

A prism is a *right prism* if and only if at least one lateral edge is perpendicular to the planes containing the bases.

A prism is a *triangular prism, quadrilateral prism, trapezoidal prism, rectangular prism,* and so on, according as its bases are triangular regions, quadrilateral regions, trapezoidal regions, rectangular regions, and so on.

A prism is a *parallelepiped* if and only if every face is a region whose boundary is a parallelogram.

A parallelepiped is a *rectangular parallelepiped* if and only if every face is a rectangular region.

A rectangular parallelepiped is a *cube* if and only if every face is a square region.

   The models in Fig. 8.5 illustrate some of the classes of prisms.

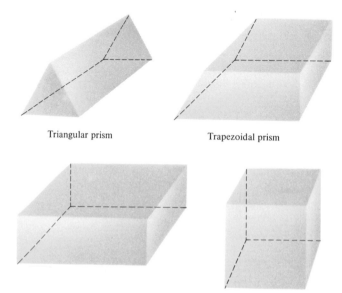

Triangular prism   Trapezoidal prism

Rectangular parallelepiped   Cube

FIG. 8.5
Models of prisms.

8.3 MEASURE OF POLYHEDRONS

There is a measure associated with polyhedrons which is analogous to the measure of a polygon; namely, the sum of the measures of the faces. The *measure of the lateral surface* of a polyhedron is the sum of the measures of the lateral faces, and the *total measure* of a polyhedron is the sum of the measure of the lateral surface and the measures of the bases.

Definition 8.9 An *altitude* of a prism is a line segment such that one endpoint is a point of a base, is perpendicular to the plane containing this base, and the other endpoint is in the plane which contains the other base.

Exercise 1 of Exercise set 8.1 gives a model which suggests that the measure of the lateral surface of a right prism $\mathscr{F}$ is ph, where p is the measure of the boundary of the base and h is the measure of an altitude of $\mathscr{F}$; that is, $m(\mathscr{F}) = ph$.

Definition 8.10 Let $\mathscr{R}$ be a polygonal region in a plane α and P a point not in α. A *pyramid* is a polyhedron which is the union of $\mathscr{R}$ and the triangular regions determined by P and consecutive vertices of $\mathscr{R}$.

In Definition 8.10, the polygonal region $\mathscr{R}$ is called the *base*, the point P the *vertex*, and each triangular region a *lateral face* of the pyramid. The *lateral surface* of a pyramid is the union of the lateral faces. The measure of the lateral surface is the sum of the measures of the lateral faces. The *total measure* is the sum of the measure of the lateral surface and the measure of the base.

Definition 8.11 The *altitude* of a pyramid is the line segment such that one endpoint is the vertex of the pyramid, the other endpoint is in the plane which contains the base, and is perpendicular to the plane of the base.

A pyramid is a *right pyramid* if and only if the boundary of the base is a regular polygon and the lateral edges are all congruent. For a right pyramid the altitudes of the lateral faces are called *slant altitudes*. It is not difficult to see that the measure of the *lateral surface* of a right pyramid $\mathscr{F}$ is $m(\mathscr{F}) = \frac{+1}{+2} ps$, where p is the measure of the boundary of the base and s is the measure o each slant altitude.

Pyramids are also classified according to their bases; a pyramic is called *triangular, square, rectangular,* and so on, according a: its base is a triangular region, square region, rectangular region and so forth. An illustration is given in Example 8.2.

EXAMPLE 8.2 The pyramid suggested by the accompanying

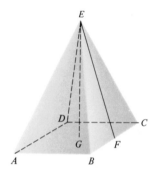

model is called a *square pyramid.* If the lateral edges $\overline{AE}$, $\overline{BE}$ $\overline{CE}$, and $\overline{DE}$ are all congruent, then the pyramid is a right square pyramid (that is, a right pyramid whose base is a square region) where $\overline{EG}$ is the altitude of the pyramid and $\overline{EF}$ is a slant altitude

Exercise set 8.1

1. Consider the accompanying model of a right trapezoidal

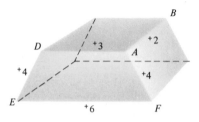

prism. The numerals indicate the measures of the edges Imagine that the lateral faces are "unfolded" at $\overline{AB}$ as sug- gested by the accompanying model.

a. What is the measure of ▱DEFA?
b. What is the measure of an altitude of the prism?
c. Determine the measure of the lateral surface from the unfolded faces.
d. Does your answer in part c agree with ph, where p is the measure of ▱DEFA and h is the measure of an altitude of the prism?

2. What is the least number of faces that a prism can have?

3. Let $ABCD$ and $EFGH$ be congruent trapezoids in parallel planes in the accompanying model.

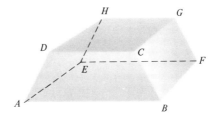

a. The model is a model of a _____ prism.
b. ▰ABCD is called a _____.
c. ▰ADHE is called a _____.
d. $\overline{AE}$ is called an _____.
e. If $\overline{AE}$ were perpendicular to the plane containing ▱ABCD, then the prism would be called a _____.

4. A right prism has a lateral edge of measure $^{+}3$ and the measure of the boundary of its base is $^{+}24$. What is the measure of its lateral surface?

5. Determine the measure of an altitude of a right prism for which the measure of the lateral surface is $^{+}480$ and the measure of the boundary of the base is $^{+}40$.

6. What is the measure of the lateral surface of a cube whose lateral edges have measure $^+5$?

7. Prove: The measure of the lateral surface of a cube whose lateral edges have measure e is $^+4e^2$.

8. Prove: The total measure of a cube whose lateral edges have measure e is $^+6e^2$.

9. Refer to the model of a rectangular parallelepiped with ■$ABCD$ and ■$EFGH$ as bases.

   a. Name the polygon which forms the boundary for each face.
   b. How many faces does the parallelepiped have?
   c. Is $\overline{EF} \parallel \overline{AB}$? Why?
   d. If $m(\overline{AB}) = {}^+4$, $m(\overline{BF}) = {}^+2$, and $m(\overline{BC}) = {}^+3$, what is the measure of the lateral surface of the parallelepiped?

10. The measure of one edge of the base of a right square pyramid is $^+10$ and the measure of a slant altitude is $^+8$. What is the measure of the lateral surface of the pyramid?

11. Determine the total measure of a right square pyramid whose slant altitude has a measure of $^+13$ and the measure of a side of the base is $^+17$.

8.4 CIRCULAR CYLINDERS, CIRCULAR CONES, AND THEIR MEASURES

Although cylinders and cones are familiar to the reader from experiences in the real world, we will give them a formal mathematical definition.

Definition 8.12 Let α and β be two parallel planes. Let $\mathscr{C}_1$ and $\mathscr{C}_2$ be two congruent circles in α and β, respectively. A *circular cylinder* is the union of the two circular regions determined by $\mathscr{C}_1$ and $\mathscr{C}_2$ and the set of all parallel line segments that have one endpoint in $\mathscr{C}_1$ and the other in $\mathscr{C}_2$.

The circular regions determined by $\mathscr{C}_1$ and $\mathscr{C}_2$ in Definition 8.12 are called *bases* of the circular cylinder. The surface determined by the union of the parallel line segments is called the *lateral surface.* A circular cylinder is a *right* circular cylinder if and only if the line segments forming the lateral surface are perpendicular to the two planes which contain the bases.

Definition 8.13 An *altitude* of a circular cylinder is a line segment which has one endpoint in a base, is perpendicular to the plane which contains this base, and has the other endpoint in the plane which contains the other base.

Some models of circular cylinders are shown in Figs. 8.6(a and b).

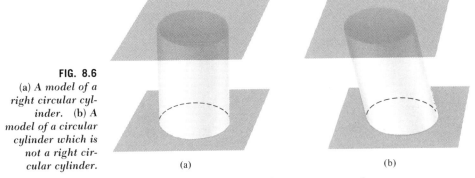

FIG. 8.6
(a) *A model of a right circular cylinder.* (b) *A model of a circular cylinder which is not a right circular cylinder.*

(a)   (b)

It is possible to define cylinders which have regions determined by various pairs of congruent simple closed curves (not union-of-line-segments curves) in parallel planes as bases. However, we will restrict our discussion to right circular cylinders. The measure of the *lateral surface* of a right circular cylinder $\mathscr{F}$ is given by $m(\mathscr{F}) = {}^+2\pi rh$, where r is the radius of the circular boundaries of the bases and h is the measure of an altitude of the circular cylinder. The models in Fig. 8.7 and the discussion which follows may give some intuition for this formula.

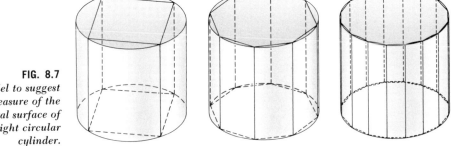

FIG. 8.7
Model to suggest the measure of the lateral surface of a right circular cylinder.

Recall that the measure of the lateral surface of a right prism is given by ph, where p is the measure of the polygonal boundary of the base and h is the measure of an altitude of the prism.   Moreover, we suggested earlier in a similar diagram that as the number of sides of the polygons "inscribed" in the circular bases as illustrated, becomes greater, the measure of the boundary of the polygonal region approaches the measure $^+2\pi r$ of the circle.   Moreover, the measure of the altitudes of the inscribed prisms is equal to the measure of an altitude of the circular cylinder.   Therefore, as the number of sides of the inscribed polygons becomes greater, ph approaches $^+2\pi rh$.

Definition 8.14 Let $\mathscr{C}$ be a circle in a plane α and P a point not in α.   A *circular cone* is the union of the circular region determined by $\mathscr{C}$ and all line segments PX such that $X \in \mathscr{C}$.

The circular region determined by $\mathscr{C}$ is called the *base* of the cone and the surface formed by the union of the line segments PX is the *lateral surface*.   The point P is called the *vertex*.

Definition 8.15 The *altitude* of a circular cone is the line segment with the vertex as one endpoint, the other endpoint in the plane containing the base, and perpendicular to the plane containing the base.

A model of a circular cone is shown in Fig. 8.8.

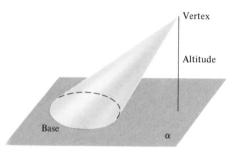

Vertex

Altitude

Base

α

FIG. 8.8
Model of a circular cone.

The reader should reflect on the definitions of circular cylinder and prism, and on the definitions of circular cone and pyramid to note the similarity.

A circular cone is a *right* circular cone if and only if one endpoint of the altitude is the center of the boundary circle of the base.

8.4 CIRCULAR CYLINDERS, CIRCULAR CONES, AND THEIR MEASURES

Every line segment PX such that P is the vertex and X is on the circle which is the boundary of the base is called a *slant altitude*.

Methods of calculus are required to determine the measure of the *lateral surface* of a right circular cone $\mathscr{F}$; we state the result without proof.

$$m(\mathscr{F}) = \frac{^+1}{^+2} \times {}^+2\pi rs = \pi rs,$$

where s is the measure of a slant altitude and r is the radius of the circular boundary of the base. The models in Fig. 8.9 and the discussion which follows may give some intuition for this formula.

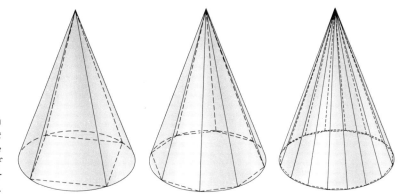

FIG. 8.9
Model to suggest the measure of the lateral surface of a right circular cone.

Recall that the measure of the lateral surface of a right pyramid is given by $\frac{^+1}{^+2} ps'$, where p is the measure of the boundary of the base and s' is the measure of a slant altitude. Moreover, we suggested earlier in a similar diagram (p. 184) that as the number of sides of the polygon "inscribed" in the circle becomes greater (as illustrated), the measure of the polygon approaches the measure $^+2\pi r$ of the circle. Also, as this happens the measure of a slant altitude of the pyramid approaches the measure of a slant altitude of the circular cone; that is, in the expression $\frac{^+1}{^+2} ps'$, p approaches $^+2\pi r$ and s' approaches s; therefore, $\frac{^+1}{^+2} ps'$ approaches $\frac{^+1}{^+2} \times {}^+2\pi r \cdot s = \pi rs$.

Exercise set 8.2

1. For a right circular cylinder, let the radii of the boundaries of the bases be r and the measure of an altitude be h. Deter-

mine the measure of the lateral surface and total measure of the right circular cylinder where r and h are, respectively:

a. $^+3$ and $^+7$ **b.** $^+6$ and $^+4$
c. $^+7$ and $^+3$ **d.** $^+2$ and $^+5$
e. $^+4$ and $^+10$ **f.** $^+6$ and $^+15$

2. If the radius of a circle is doubled, how does its measure change?

3. If the radii of the boundaries of the bases and the measure of an altitude of a right circular cylinder are doubled, how does the measure of the lateral surface and total measure change? What if each is tripled?

4. For a right circular cone, let the radius of the boundary of the base be r and the measure of a slant altitude be s. Determine the measure of the lateral surface and total measure of a right circular cone where r and s are, respectively:

a. $^+2$ and $^+5$ **b.** $^+4$ and $^+10$ **c.** $^+6$ and $^+15$

5. If the radius of the boundary of the base and measure of a slant altitude of a right circular cone are doubled, how does the measure of the lateral surface and total measure change? What if each is tripled?

6. Refer to the accompanying models which suggest the intersection of circular cylinders and circular cones with planes. Sketch a model of the intersection.

a.   **b.**

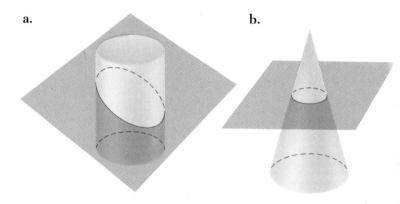

c.   **d.**

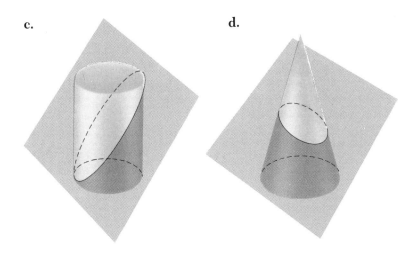

7. Is a circular cone a convex surface?

8. Is a circular cylinder a convex surface?

9. Give a definition of the interior of a circular cone.

10. Give a definition of the interior of a circular cylinder.

8.5 SOLIDS, MEASURE OF SOLIDS

8.5.1 Introduction

A *solid* is a set of points which is the union of a simple closed surface and its interior. If $\mathscr{F}$ is a simple closed surface, the solid which is the union of $\mathscr{F}$ and its interior we will refer to as *solid $\mathscr{F}$*. For example, a solid determined by a cube (that is, the union of a cube and its interior) we will call a *solid cube*.

To approximate the measure of model line segments, model angles, and model regions we chose models of unit-line segments, angles, and regions, respectively. To approximate the measure of model solids we proceed analogously; that is, we choose a model unit-solid and compare a given model solid to the unit. The consideration of the measure of model solids will provide motivation for the determination of the exact measure of geometric solids.

Since a cube is to space as a square is to a plane, we choose as a unit-solid a solid cube which has edges with measure $^+1$ with respect to some unit-pair. We associate with the unit-solid cube the measure $^+1$.

Through a sequence of models we will illustrate the procedure by which we approximate the measure of a model solid rectangular parallelepiped. We let the accompanying be our model of the solid unit-cube. The measure of the solid unit-cube is $^+1$. If we

 ← model solid unit-cube

place four model solid unit-cubes face to face, as in the accompanying model, it is clear that the measure, with respect to the

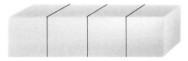

model solid unit-cube, of the solid which is the union of these four solids is $^+4$. Now if we place three such rows of model solid unit-cubes face to face, as in the accompanying model, the measure,

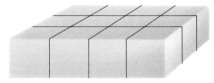

with respect to the model solid unit-cube, of the resulting model solid is $^+12$ (that is, $^+3 \times {}^+4$).

Now if we stack two of these model solids, as in the accompanying model, the measure of the resulting solid, with respect to

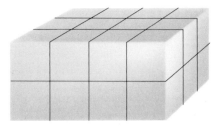

the model solid unit-cube, is $^+24$ (that is, $^+2 \times {}^+12$). We call the reader's attention to the fact that $^+12$ is also the measure of the boundary of the base of the parallelepiped which resulted from this joining of model solid unit-cubes. Moreover, the measure of this base is given with respect to a model square unit whose sides have measure $^+1$ with respect to our chosen unit-square.

In the above illustration we chose a model solid unit-cube and then built a model solid parallelepiped using more models of the solid unit-cube. In this way we were able to determine quite accurately the measure of the resulting solid parallelepiped. However, it is not always possible to be quite this accurate, as illustrated in Example 8.3.

EXAMPLE 8.3 The following is suggestive of the approximation of the measure of the solid parallelepiped with respect to the accompanying model solid unit-cube.

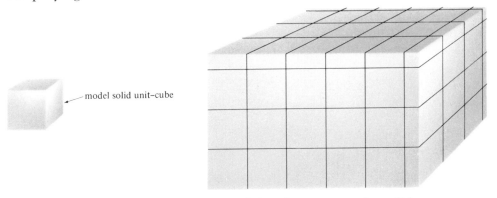

model solid unit–cube

We conclude that the measure of the solid with respect to the solid unit-cube is greater than $^+45$ and less than $^+96$.

The word "volume" is to solids and measures of solids as the word "area" is to regions and measures of regions, as "length" is to line segments and measures of line segments, and so on. That is, a measure of a solid is a positive real number which is determined by comparing the given solid to a unit solid; the *volume* of a solid is a measure of the solid together with the unit used to determine the measure. We frequently use a solid cube whose edges have an inch measure of $^+1$ as a unit solid; such a cube is called an *inch cube* or *cubic inch*; its measure is $^+1$ and we say its volume is $^+1$ cubic inch. If we determined that the cubic inch measure of a solid was $^+3 \cdot {}^+4$, then we would report that its volume is $^+12$ cubic inches.

8.5.2 Measure of a solid rectangular parallelepiped

We start out by making sure that we have a measure function which assigns to every solid a positive real number; this we guarantee in Postulate 8.2.

Postulate 8.2 (*Solid measure postulate*) There exists a measure function *m* which assigns to every solid a unique positive real number relative to a solid unit-cube, which is assigned $^+1$. If $\mathscr{S}$ denotes any solid, then the real number $m(\mathscr{S})$ is called the *measure* of $\mathscr{S}$.

We hope the illustrations in the previous section make the following assumption seem reasonable.

Postulate 8.3 For every solid parallelepiped $\mathscr{S}$, the measure of $\mathscr{S}$, denoted by $m(\mathscr{S})$, is given by $m(\mathscr{S}) = bh$, where b is the measure of the base and h is the measure of an altitude of the parallelepiped.

Postulate 8.3 is illustrated in Example 8.4.

EXAMPLE 8.4 Consider the solid rectangular parallelepiped

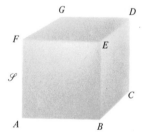

$\mathscr{S}$ suggested by the accompanying model. If $m(\overline{AB}) = m(\overline{BC}) = {}^+3$ and $m(\overline{FA}) = {}^+7$, then $m(\mathscr{S}) = ({}^+3 \times {}^+3) \times {}^+7$.

Exercise set 8.3

1. Refer to the model of $\mathscr{S}$ in Example 8.4. Determine $m(\mathscr{S})$ in each of the following cases:

   a. $m(\overline{FA}) = {}^+4$, $m(\overline{AB}) = {}^+3$, and $m(\overline{BC}) = {}^+2$.
   b. $m(\overline{FA}) = {}^+8$, $m(\overline{AB}) = {}^+3$, and $m(\overline{BC}) = {}^+2$.
   c. $m(\overline{FA}) = {}^+8$, $m(\overline{AB}) = {}^+6$, and $m(\overline{BC}) = {}^+2$.
   d. $m(\overline{FA}) = {}^+8$, $m(\overline{AB}) = {}^+6$, and $m(\overline{BC}) = {}^+4$.

2. Compare your results in parts a and b of Exercise 1. What do you observe?

3. Compare your results in parts a and c of Exercise 1. What do you observe?

4. Compare your results in parts a and d of Exercise 1. What do you observe?

5. Refer to the model of $\mathscr{S}$ in Example 8.4.

   a. If $m(\overline{FA}) = m(\overline{AB}) = m(\overline{BC}) = {}^+2$, then $m(\mathscr{S}) =$ _____.
   b. If $m(\overline{FA}) = m(\overline{AB}) = m(\overline{BC}) = {}^+5$, then $m(\mathscr{S}) =$ _____.

6. Prove that the measure of a cubical solid is e^3, where e is the measure of each of the edges.

7. If the measure of the edge of a cubic solid is doubled, how does its measure change?

8. Determine the measure of the edge of a cubic solid if its measure is equal to the measure of its lateral surface.

9. Draw models of the intersection of the solids determined by rectangular parallelepipeds and planes as suggested in the accompanying models.

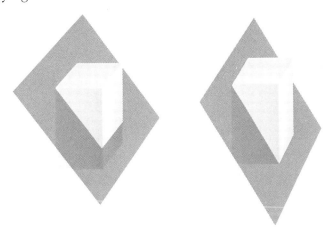

8.5.3 Measure of solid right prisms

We will proceed to determine the measure of certain solid prisms from our knowledge of the measure of solid rectangular parallelepipeds. Our procedure parallels the development of the measure of right triangular regions from the measure of rectangular regions. To accomplish this we need some definitions and postulates.

Definition 8.16 Two right prisms are *congruent* if and only if there exists a one-to-one correspondence between the vertices such that corresponding angles and edges are congruent.

Postulate 8.4 If two prisms are congruent, then the solids determined by them have equal measures.

Postulate 8.5 If two solid prisms $\mathscr{S}_1$ and $\mathscr{S}_2$ determined by prisms $\mathscr{P}_1$ and $\mathscr{P}_2$ are such that $\text{Int}(\mathscr{P}_1) \cap \text{Int}(\mathscr{P}_2) = \varnothing$, then $m(\mathscr{S}_1 \cup \mathscr{S}_2) = m(\mathscr{S}_1) + m(\mathscr{S}_2)$.

Let $\mathscr{S}_1$ (see Fig. 8.10) be a solid rectangular parallelepiped; a plane separates $\mathscr{S}_1$ into two solid right triangular prisms, $\mathscr{S}_2$ and

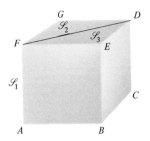

FIG. 8.10
Separation of a solid by a plane.

$\mathscr{S}_3$. We leave to the reader the problem of convincing himself that the one-to-one correspondence defined by the pairing

$G \leftrightarrow E$, $H \leftrightarrow B$ (H is the "hidden" vertex in the figure),

$F \leftrightarrow D$, $A \leftrightarrow C$,

$D \leftrightarrow F$, $C \leftrightarrow A$

is a congruence between the two right prisms; therefore,

$$m(\mathscr{S}_2) = m(\mathscr{S}_3).$$

Moreover, by Postulate 8.5,

$$m(\mathscr{S}_1) = m(\mathscr{S}_2) + m(\mathscr{S}_3),$$

$$m(\mathscr{S}_2) = \frac{{}^{+}1}{{}^{+}2} \times m(\mathscr{S}_1) = \frac{{}^{+}1}{{}^{+}2}\, bh.$$

Now, $b = m(\blacksquare ABCH)$ and hence $\dfrac{^{+}1}{^{+}2}\, b = m(\blacktriangle AHC)$.   Therefore, the measure of the solid determined by the right triangular prism $\mathscr{S}_2$ is given by $m(\mathscr{S}_2) = b'h$, where b' is the measure of a base which is a triangular region and h is the measure of the altitude of the prism.

Since very solid right prism can be separated into solid right triangular prisms (see Fig. 8.11), it is possible to show that the mea-

FIG. 8.11
*A solid right prism
separated into
right triangular
prisms.*

sure of every solid right prism $\mathscr{S}$, is given by $m(\mathscr{S}) = bh$, where b is the measure of a base which is a polygonal region and h is the measure of an altitude of the prism.

In fact, by using geometric theorems, which space limitations prohibit us from pursuing, it is possible to show that the measure of every solid $\mathscr{S}$ determined by a prism is given by $m(\mathscr{S}) = bh$, where b is the measure of a base of the prism and h is the measure of an altitude of the prism.

8.5.4 Measure of other solids

We conclude our study of the measure of solids by simply stating the formulas to determine the measure of several common types of solids.

The measure of a solid right circular cylinder $\mathscr{S}$, is $m(\mathscr{S}) = \pi r^2 h$, where r is the radius of the circular boundary of the base and h is the measure of an altitude.

The measure of a solid pyramid $\mathscr{S}$, is $m(\mathscr{S}) = \dfrac{^{+}1}{^{+}3}\, bh$, where b is the measure of the base which is a polygonal region and h is the measure of the altitude.

The measure of a solid right circular cone $\mathscr{S}$, is $m(\mathscr{S}) = \dfrac{^{+}1}{^{+}3}\pi r^2 h$, where r is the radius of the circular boundary of the base and h is the measure of the altitude.

The measure of a solid sphere $\mathscr{S}$, is $m(\mathscr{S}) = \dfrac{^{+}4}{^{+}3}\pi r^3$, where r is the radius of the sphere.

Exercise set 8.4

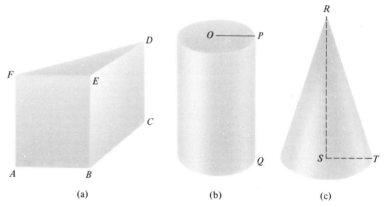

(a)   (b)   (c)

1. Refer to the solid right triangular prism suggested by accompanying model (a). Determine the measure when:

   a. $\triangle FED$ is a right triangle with right angle FED, $m(\overline{FE}) = {}^{+}4$, $m(\overline{ED}) = {}^{+}3$, and $m(\overline{EB}) = {}^{+}7$.

   b. $m(\overline{AC}) = {}^{+}5$, the measure of the altitude of $\triangle ABC$ from B to $\overline{AC}$ is ${}^{+}3$, and $m(\overline{FA}) = {}^{+}12$.

2. Consider a solid right triangular pyramid whose base is $\triangle XYZ$ and vertex is W. Determine its measure when:

   a. $\triangle XYZ$ is a right triangle with right angle XYZ, $m(\overline{XY}) = m(\overline{YZ}) = {}^{+}5$, and $m(\overline{YW}) = {}^{+}6$, where $\overline{YW}$ is the altitude of the pyramid. Draw a model of the pyramid.

   b. $m(\blacktriangle XYZ) = {}^{+}12$, and $m(\overline{YW}) = {}^{+}4$, where $\overline{YW}$ is the altitude of the pyramid.

3. Consider the solid right circular cylinder suggested by accompanying model (b). Determine its measure when:

a. $m(\overline{OP}) = {}^+3$ and $m(\overline{PQ}) = {}^+4$.
b. $m(\overline{OP}) = {}^+6$ and $m(\overline{PQ}) = {}^+4$.
c. $m(\overline{OP}) = {}^+9$ and $m(\overline{PQ}) = {}^+4$.
d. What generalization do you conjecture from parts a–c?
e. $m(\overline{OP}) = {}^+3$ and $m(\overline{PQ}) = {}^+8$.
f. $m(\overline{OP}) = {}^+3$ and $m(\overline{PQ}) = {}^+12$.
g. What generalization do you conjecture from parts a, e, and f?

4. Consider the solid right circular cone suggested by accompanying model (c). Determine its measure when:

a. $m(\overline{RS}) = {}^+3$ and $m(\overline{ST}) = {}^+2$.
b. $m(\overline{RS}) = {}^+6$ and $m(\overline{ST}) = {}^+2$.
c. $m(\overline{RS}) = {}^+9$ and $m(\overline{ST}) = {}^+2$.
d. What generalization do you conjecture from parts a–c?
e. $m(\overline{RS}) = {}^+3$ and $m(\overline{ST}) = {}^+4$.
f. $m(\overline{RS}) = {}^+3$ and $m(\overline{ST}) = {}^+8$.
g. What generalization do you conjecture from parts a, e, and f?

5. Complete the accompanying table to discover a relationship between the total measure of a cube and the measure of the solid determined by the cube.

Measure of each edge of the cube	M_1: total measure of the cube	M_2: measure of solid cube	Ratio: $\dfrac{M_1}{M_2}$
$^+1$	$^+6$	$^+1$	$^+6$
$^+2$	$^+24$	$^+8$	$^+3$
$^+3$	$^+54$	$^+27$	$^+2$
$^+4$			
$^+5$			
$^+6$			
$\vdots$			
x			

6. Complete the table on the next page to discover a relationship between the measure of sphere and the solid determined by the sphere.

8   subsets of space—surfaces and solids

Radius of the sphere	M_1: measure of the sphere[a]	M_2: measure of the spherical solid	Ratio: $\dfrac{M_1}{M_2}$
+1			
+2			
+3			
+4			
+5			
+6			
.			
.			
.			
x			

[a] *Note:* The measure of a sphere $\mathscr{S}$ is given by $m(\mathscr{S}) = {}^+4\pi r^2$.

8.6 IN RETROSPECT

We will give a brief overview of our study of geometry by return-ing to the two objectives we stated in Chapter 2.   One objective was to demonstrate the structure of mathematics in the setting of geometry.   To this end we pursued for a short time a formal de-ductive development of Euclidean geometry proving every desired result from the basic postulates.   This introduction should suggest the spirit, necessary considerations, and the magnitude of the task of undertaking to develop the whole system of Euclidean geometry in this manner.

After this formal introduction we proceeded less formally in terms of the degree of detail in the proofs; however, the entire de-velopment still is in the spirit of *developing* the geometric con-cepts so that the *developmental nature* of mathematics was in evidence.   To call attention to this, we single out a sequence of concepts which developed one from another starting with an *un-defined* term.

The undefined concept of *line* together with postulated and proved characteristics of lines was used to define *betweenness*, which in turn was used in defining *line segment*, which in turn was used to define *convex set* (and other concepts), which in turn was used to define other concepts.

Our second objective was to present quite accurately many geometric concepts and the vocabulary with which prospective teachers should be familiar.

In the sequence we made several remarks concerning a distinction between: a measure and the length of a line segment, a measure and the area of a region, and a measure and the volume of a solid, and so on. *Measurement* is an all inclusive word of which length, perimeter, circumference, area, and volume are special cases. Every geometric figure has many measures; each measure of a geometric figure is a real number which is determined by comparing the given geometric figure to some unit of measure. Every geometric figure has exactly one measurement; the measurement of a geometric figure is given by indicating the measure and the unit. The words which denote the measurement of certain geometric figures are given in Table 8.1.

TABLE 8.1 Words used for measurements of geometric figures

Geometric figure	*Word for its measurement*
Line segment	Length
Angle	
Circle	Circumference
Polygon	Perimeter
Region	Area
Surface	Area
Solid	Volume

operations, groups, transformations, transformation groups

In Chapter 10 we shall present an approach to Euclidean geometry which is frequently referred to as *transformation geometry*. To study this approach to geometry it is necessary to be familiar with the concepts of operation, group, transformation, and transformation group; these concepts constitute the topics of this chapter.

9.1 OPERATIONS

The word "operation" is familiar to you from arithmetic; we speak of the operation of addition, and so on. If we think further about the operation of addition, we can easily observe that addition is a rule (one we have memorized) which assigns to *pairs of numbers* a unique number. Now, such an assignment defines a function!

Examples 9.1 and 9.2 give illustrations to further emphasize that operations are particular types of functions.

EXAMPLE 9.1 Let $C = \{1,2,3,4,5, \ldots\}$. The following function from $C \times C$ to C is called "addition of counting numbers."

$$F = \{((1,1),2),((1,2),3),((2,1),3),((1,3),4),((2,2),4),((3,1),4),$$
$$((1,4),5), \ldots\}.$$

The accompanying is a partial mapping diagram of the function "addition of counting numbers."

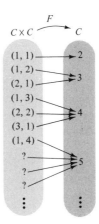

EXAMPLE 9.2 Let $S = \{0,1,2\}$. The following function $\oplus$ is an operation on S.

$$\oplus = \{((0,0),0),((0,1),1),((0,2),2),((1,0),1),((1,1),2),((1,2),0),$$
$$((2,0),2),((2,1),0),((2,2),1)\}.$$

Note that $\oplus$ is a function from $S \times S$ to S. That is, $(0,0) \overset{\oplus}{\to} 0$, $(0,1) \overset{\oplus}{\to} 1$, $(0,2) \overset{\oplus}{\to} 2$, $(1,0) \overset{\oplus}{\to} 1$, and so on.

A precise definition of binary operation is given in Definition 9.1.

Definition 9.1 A *binary operation* $*$ on a set S is a function from $S \times S$ to S.

In view of the kind of questions about which we will concern ourselves regarding operations, Definition 9.1 is not a very usable one; to arrive at a more usable definition we enumerate the properties which a relation $*$ must satisfy in order to be a function from $S \times S$ to S (that is, an operation on S).

1. For every $(x,y) \in S \times S$ there exists $z \in S$ such that $((x,y),z) \in *$.
2. For all (x,z), $(y,w) \in S \times S$ and for all p, $q \in S$, if $((x,z),q) \in *$ and $((y,w),p) \in *$ and $(x,z) = (y,w)$, then $q = p$.

If $*$ is a binary operation on a set S, the element which $*$ assigns to the ordered pair (x,y) is denoted by $x * y$; that is: for every (x,y) $\in S \times S$, $(x,y) \overset{*}{\to} x * y$.   Using this notation, we enumerate again in Definition 9.2 the properties that a relation $*$ must satisfy to be an operation on a set S.   It is this definition which we will find most helpful in answering questions about whether or not certain relations are binary operations and in other applications of the concept of binary operation.

Definition 9.2   A *binary operation* $*$ on a set S is a relation which assigns to each ordered pair $(x,y) \in S \times S$ an element $x * y$, such that:

1.   For all $x, y \in S, X * y \in S$.   The Closure Property.
2.   For all $x, y, z, w \in S$, if $x = y$ and $z = w$, then $x * z = y * w$. The Well Defined Property.

(If $*$ satisfies property 1, we say that S is closed with respect to $*$.)

The following intuitive summary may be helpful.   A binary operation $*$ on a set S is a rule which:

1.   Assigns to *every* ordered pair in $S \times S$ an element of S.
2.   To every ordered pair in $S \times S$, *only one* element of S is assigned.

We have referred to union as an operation on sets.   This terminology is appropriate since the definition of union does give a rule by which a unique set is assigned to every ordered pair of sets.   A special technique for defining a relation is illustrated in Examples 9.3 and 9.4.

EXAMPLE 9.3   Let $S = \{0,1,2\}$ and let $R = \{((0,0),0),((0,1),1),$ $((0,2),2),((1,0),1),((1,1),2),((1,2),3),((2,0),2),((2,1),3),((2,2),4)\}$ be a relation whose domain is $S \times S$.   A table conveniently defines this relation.

R	0	①	2
0	0	1	2
1	1	2	3
②	2	③	4

To determine, from the table, the element that the relation R assigns to, for example, the ordered pair (2,1), proceed as follows:

1. "Spot" 2 in the left most column ($\square$).
2. "Spot" 1 in the very top row ($\bigcirc$).
3. The element assigned to (2,1) is 3 ($\triangle$).

From the table it is easy to see that S is not closed with respect to R. Therefore, R is not an operation on S.

EXAMPLE 9.4 The following table defines a binary operation $\otimes$ on $S = \{0,1\}$.

$\otimes$	0	1
0	0	0
1	0	1

From the table we determine that

$$0 \otimes 0 = 0, \quad 0 \otimes 1 = 0, \quad 1 \otimes 0 = 0, \quad \text{and} \quad 1 \otimes 1 = 1;$$

that is,

$$(0,0) \xrightarrow{\otimes} 0, \ (0,1) \xrightarrow{\otimes} 0, \ (1,0) \xrightarrow{\otimes} 0, \text{ and } (1,1) \xrightarrow{\otimes} 1.$$

If $*$ is a binary operation on a set S and x, y, and z are elements of S, then $x * y * z$ is not defined. We remove this difficulty for all operations in Definition 9.3.

Definition 9.3 Let $*$ be a binary operation on a set S, then for all $x, y, z \in S, x * y * z = (x * y) * z$.

[*Note:* The decision to define $x * y * z$ to be $(x * y) * z$ as opposed to $x * (y * z)$ is quite arbitrary. However, to avoid possible ambiguity, a choice must be made; we chose as in Definition 9.3.]

We will be interested in certain properties which some operations satisfy. The reader is no doubt familiar with the terms "commutative," "associative," and others as they apply to operations on numbers. We give a general definition of these words in Definition 9.4.

Definition 9.4 If $*$ and $\#$ are binary operations on a set S, then:

1. $*$ is *associative* if and only if

    for all x, y, $z \in S$, $(x * y) * z = x * (y * z)$.

2. $*$ is *commutative* if and only if

    for all x, $y \in S$, $x * y = y * x$.

3. $*$ is said to be *distributive over $\#$ from the left* if and only if

    for all x, y, $z \in S$, $x * (y \# z) = (x * y) \# (x * z)$.

4. $*$ is said to be *distributive over $\#$ from the right* if and only if

    for all x, y, $z \in S$, $(x \# y) * z = (x * z) \# (y * z)$.

5. An element $e \in S$ is called an *identity* for $*$ if and only if

    for every $x \in S$, $x * e = x$ and $e * x = x$.

6. Let $e \in S$ be the identity for $*$. For all x, $y \in S$, x is called an *inverse* of y with respect to $*$ if and only if

    $x * y = e$ and $y * x = e$.

Exercise set 9.1

1. Let $Q = \{a,b,c,d,e\}$ and $*$ be the relation from $Q \times Q$ to Q defined by the table.

$*$	a	b	c	d	e
a	a	b	c	d	e
b	b	c	a	e	c
c	c	a	b	b	a
d	b	e	b	e	d
e	d	b	a	d	d

    a. Is Q closed with respect to $*$?
    b. Is $*$ well defined on Q?
    c. Is $*$ a binary operation on Q?

d. Compute the following:
   (1) $(a * b) * (b * d)$,   (2) $[(a * b) * e] * d$,
   (3) $[a * (b * e)] * (b * c)$.

e. Compute $(a * b) * c$ and $a * (b * c)$. Can you determine from this whether $*$ is associative?

f. Compute $(a * c) * d$ and $a * (c * d)$. Can you determine from this whether $*$ is associative?

g. Compute $(b * c) * d$ and $b * (c * d)$. Can you determine from this whether $*$ is associative?

h. Is $*$ commutative?

i. Determine the replacement(s) for x from Q which makes each of the following a true statement:
   (1) $a * x = c$.   (2) $(a * x) * d = b$.
   (3) $c * (x * e) = b$.

2. Let $Q = \{a,b,c\}$ and $\boxdot$ be the relation from $Q \times Q$ to Q defined as follows:

$(a,a) \xrightarrow{\boxdot} a$ $(b,a) \xrightarrow{\boxdot} b$ $(c,a) \xrightarrow{\boxdot} c$

$(a,b) \xrightarrow{\boxdot} a$ $(b,b) \xrightarrow{\boxdot} b$ $(c,b) \xrightarrow{\boxdot} c$

$(a,c) \xrightarrow{\boxdot} a$ $(b,c) \xrightarrow{\boxdot} b$ $(c,c) \xrightarrow{\boxdot} c$

$(a,b) \xrightarrow{\boxdot} b$

a. Is Q closed with respect to $\boxdot$?

b. Is $\boxdot$ well defined on Q?

c. Is $\boxdot$ a binary operation on Q?

d. Is $\boxdot$ commutative?

e. Is $\boxdot$ associative?

3. Complete the following tables so they define commutative binary operations on $S_1 = \{a,b\}$, $S_2 = \{a,b,c\}$, and $S_3 = \{a,b,c,d\}$.

a.

$\cdot$	a	b
a	a	
b	b	a

b.

$\odot$	a	b	c
a	a		b
b	c	a	
c		c	a

c.

$\circledcirc$	a	b	c	d
a	a	b		
b		d	a	c
c	c		d	
d	d		b	a

4. Let $S = \{1,2,3\}$.   Define $\circledast$ on S as follows:

For all $x, y \in S, x * y = x + y.$

For example:

$1 \circledast 3 = 1 + 3 = 4.$

a.   Complete the following table for $\circledast$.

$\circledast$	1	2	3
1			
2			
3			

b.   Is $\circledast$ well defined on S?
c.   Is S closed with respect to $\circledast$?
d.   Is $\circledast$ a binary operation on S?

5.   Let $C_e = \{2,4,6,8\}$.   Define $\boxast$ on C_e as follows:

For all $x, y \in C_e, x \boxast y = x + y.$

For example:

$4 \boxast 6 = 4 + 6 = 10.$

a.   Is $\boxast$ well defined on C_e?
b.   Is C_e closed with respect to $\boxast$?
c.   Is $\boxast$ a binary operation on C_e?

6.   Let $C_o = \{1,3,5,7, \ldots\}$.   Define # on C_o as follows:

For all $x, y \in C_o, x \text{ \# } y = x + y.$

a.   Is # well defined on C_o?
b.   Is C_o closed with respect to #?
c.   Is # a binary operation on C_o?

7.   Let $S = \{0,1,2,3\}$.   Let $\oplus$ be defined on S as follows:

For all $x, y \in S, x \oplus y = x + y$ if $x + y < 4,$
$$x \oplus y = (x + y) - 4 \text{ if } x + y \geqslant 4.$$

For example:

$1 \oplus 2 = 3$ since $1 + 2 = 3$ and $3 < 4,$
$2 \oplus 3 = 1$ since $2 + 3 = 5$ and $5 - 4 = 1.$

a. Complete the following table for ⊕.

⊕	0	1	2	3
0				
1				
2				
3				

b. Is S closed with respect to ⊕?
c. Is ⊕ well defined on S?
d. Is ⊕ a binary operation on S?
e. Is ⊕ commutative?
f. Define • to be a binary operation on S as follows:

For all $x, y \in S$, $x \cdot y = x \cdot y$ if $x \cdot y < 4$,
$$x \cdot y = x \cdot y - 4 \text{ if } x \cdot y \geqslant 4 \text{ and } x \cdot y < 8,$$
$$x \cdot y = 1 \text{ if } x = 3 \text{ and } y = 3.$$

For example:

$2 \cdot 1 = 2$, $2 \cdot 3 = 6 - 4 = 2$, $3 \cdot 3 = 1$.

Complete an operation table for •.
g. Is • commutative?
h. Compute $2 \cdot (1 ⊕ 3)$ and $(2 \cdot 1) ⊕ (2 \cdot 3)$. Can you determine from this whether • distributes over ⊕?
i. Compute $3 ⊕ (2 \cdot 1)$ and $(3 ⊕ 2) \cdot (3 ⊕ 1)$. Can you determine from this whether ⊕ distributes over •?

8. Let ⊙ be a binary operation defined on $S = \{a,b,c,d\}$ by the following table.

⊙	a	b	c	d
a	c	a	d	b
b	a	b	c	d
c	d	c	b	a
d	b	d	a	c

a. Does S contain an identity element with respect to $\odot$? If yes, what is it?

b. What are the inverses of a, b, c, and d with respect to $\odot$?

9. Let $C = \{1,2,3,4,5, \ldots\}$, and $*$ and $\#$ be binary operations on C which have the following properties:

$*$ and $\#$ are associative.

$*$ is commutative, but $\#$ is not.

$*$ distributes over $\#$, both from the left and the right.

$\#$ does not distribute over $*$.

Indicate whether the statements below are true or false. If you say true, indicate the property of the operation(s) which justifies it; if you say false, explain.

For example:

$(1 * 2) * 3 = 1 * (2 * 3)$. T—the associative property of $*$.

$(1 * 2) * 3 = 3 * (1 * 2)$. T—closure property of $*$ and commutative property of $*$.

a. $1 * 2 = 2 * 1$.

b. $1 * (2 * 3) = (1 * 2) * 3$.

c. $1 \# (2 \# 3) = (1 \# 2) \# 3$.

d. For all $x, y, z \in S$, $(x * y) * z = x * (y * z)$.

e. For all $x, y, z \in S$, $(x \# y) \# z = x \# (y \# z)$.

f. $(1 * 2) \# (1 * 3) = 1 * (2 \# 3)$.

g. For all $a, b, c \in S$, $(a * b) \# c = (a \# c) * (b \# c)$.

h. $(1 * 2) \# 3 = 3 \# (1 * 2)$.

i. $(1 \# 2) * 3 = 3 * (1 \# 2)$.

j. $(1 * 2) \# (3 \# 4) = (3 \# 4) \# (1 * 2)$.

9.2 DEFINITION OF GROUP

The reader is familiar with several properties which are enjoyed by the operation of addition of integers. In particular, the following are important properties which hold for addition of integers.

G_1. For all x, y, and z which are integers, $(x + y) + z = x + (y + z)$. $+$ is associative.

G_2. There is the integer 0 such that for every integer x, $x + 0 = x$ and $0 + x = x$. 0 is the identity with respect to $+$.

G₃. For every integer x there exists a unique integer w such that $x + w = 0$ and $w + x = 0$. w is the inverse of x with respect to $+$.

Since addition of integers enjoys properties G_1, G_2, and G_3, we say the set of integers together with addition forms a *group*. More generally, a *group* is a mathematical system consisting of a nonempty set and a binary operation on the set such that properties G_1, G_2, and G_3 are satisfied. The formal definition of group is given in Definition 9.5.

Definition 9.5 A *group* is a mathematical system which consists of a nonempty set G and a binary operation $*$ on G such that:

G₁. For all $x, y, z \in G$, $(x * y) * z = x * (y * z)$. $*$ is *associative*.

G₂. There exists $e \in G$ such that for every $x \in G$, $x * e = x$ and $e * x = x$. e is an *identity with respect to* $*$.

G₃. For every $x \in G$, there exists $w \in G$ such that $x * w = e$ and $w * x = e$. w is an *inverse of x with respect to* $*$.

We denote the system by $G, *$. The group $G, *$ is called a *commutative group* or *noncommutative group* according as $*$ is or is not commutative. The exercises in Exercise Set 9.2 present some problems and questions which illustrate further the concepts related to groups.

Exercise set 9.2

1. Let $G = \{a, b, c\}$ and $\bullet$ be a binary operation defined on G by the following table.

$\bullet$	a	b	c
a	a	b	c
b	b	c	a
c	c	a	b

    a. Is there an identity with respect to $\bullet$? Prove your answer.
    b. Does every element of G have an inverse with respect to $\bullet$? Prove your answer.

c.   Assuming that • is associative, is *G*, • a group?

d.   Is *G*, • a commutative group?   Prove your answer.

2.   Let *H* = {*a,b,c,d*} and # be the operation on *H* defined by the following table.

#	*a*	*b*	*c*	*d*
a	*d*	*b*	*a*	*b*
b	*b*	*a*	*b*	*c*
c	*a*	*b*	*c*	*d*
d	*a*	*c*	*d*	*b*

a.   Is there an identity with respect to #?   Prove your answer.

b.   Does every element of *H* have an inverse with respect to #?   Prove your answer.

c.   Is *H*, # a group?   Explain your answer.

3.   Clock arithmetic: Let *K* = {1,2,3,4,5,6,7,8,9,10,11,12}.   Think of the elements of *K* as arranged on the face of a clock as in the accompanying model.   Define a relation ⊛ on *K* × *K* as follows:

For all *x*, *y* ∈ *K*, *x* ⊛ *y* is the number which indicates the time *y* hours after *x* o-clock.

For example:

3 ⊛ 4 = 7 and 7 ⊛ 9 = 4.

a.   Make a table for ⊛.

b.   Is *K* closed with respect to ⊛?

c.   Is ⊛ well defined on *K*?

d.   Is ⊛ a binary operation on *K*?

e.   Prove that there is an identity with respect to ⊛.

f.   Prove that every element of *K* has an inverse with respect to ⊛.

   g. Is ⊛ associative?
   h. Is K, ⊛ a group?
   i. Is K, ⊛ a commutative group?

4. Let $K = \{0,1,2,3,4\}$ and ⊛ be defined on K as in Exercise 3. Answer parts a–i of Exercise 3.

9.3 FUNCTIONS AND GROUPS

We discussed in Chapter 1 the concepts of a one-to-one relation and a one-to-one function. We give a brief review of the concept of one-to-one function here: Recall that a function f from A to B is *one-to-one* provided that no two distinct elements of A are mapped to the same element of B. Now, according to the definition of function there *can be* some elements of B which do not have an element of A mapped to them; in this case we say A is a function from A *into* B. If every element of B has an element of A mapped to it by a function, we say the function is from A *onto* B. Note that every function from A to B which is a function from A *onto* B is also a function from A *into* B. These ideas are illustrated in Example 9.5.

EXAMPLE 9.5 Let $A = \{a,b,c,d\}$, $B = \{1,2,3,4\}$, and $C = \{1,2,3,4,5\}$. Mapping diagram (a) is a diagram of a function which

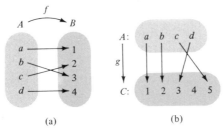

(a)   (b)

is one-to-one from A *onto* B. In diagram (b), the set C contains the element 4 which does not have an element of A mapped to it; therefore, g is not a function from A *onto* C; it is a function from A *into* C.

Our concern in this section will be with functions which are one-to-one and onto from sets to themselves. Such functions are called *transformations*.

Definition 9.6 For every set A, a function f on A is a *transformation* on A (or a *transformation* of A) if and only if f is one-to-one from A onto A.

Some transformations on a finite set are diagrammed in Example 9.6.

EXAMPLE 9.6 Let $A = \{1,2,3,4\}$. The accompanying are all diagrams of transformations on A.

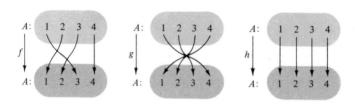

In Example 9.6 above, the function f maps 4 to itself. We say that f leaves 4 *fixed*, or that 4 is a *fixed element* of f. We leave to the reader the problem of determining the fixed elements of g and of h. Every element of the subset $\{1,2,3\}$ of A gets mapped to an element of $\{1,2,3\}$ by f. We call $\{1,2,3\}$ a *fixed set* of f. The reader should answer the question of whether $\{1,2,3\}$ is a fixed set of g or of h. Notice that none of the elements of a *fixed set* need be *fixed elements*; on the other hand, if every element of a subset of the domain of a transformation is left fixed, then the set is a fixed set.

Definition 9.7 Let t be a transformation on a set A:

a. For every $x \in A$, x is a *fixed element* of t if and only if $t(x) = x$.
b. For every $B \subseteq A$, B is a *fixed set* of t if and only if for every $x \in B$, $t(x) \in B$.

Exercise set 9.3

1. Indicate which of the following apply to the accompanying mapping diagrams.

relation	one-to-one	into
function	onto	transformation

a.

b.

c.

2. Let $A = \{1,2,3,4\}$. Indicate which of the following apply to the functions from A to A given below.

    into   one-to-one

    onto   transformation

  a. $F_1 = \{(1,2),(2,3),(3,4),(4,1)\}$.

  b. $F_2 = \{(1,1),(2,2),(3,4),(4,3)\}$.

  c. $F_3 = \{(1,3),(2,3),(3,2),(4,2)\}$.

  d. $F_4 = \{(1,4),(4,1),(3,2),(2,3)\}$.

3. Let $D = \{4,5,6,7\}$; $t = \{(4,4),(5,7),(6,6),(7,5)\}$ is a transformation of D.

  a. $t(4) = ?$, $t(5) = ?$, $t(6) = ?$, $t(7) = ?$.

  b. What are the fixed elements of t?

  c. Give a fixed set of t which contains *no* fixed elements.

  d. Give two three-element fixed sets of t each of which contains exactly one fixed element.

  e. Give a two-element fixed set of t such that both elements are fixed elements.

  f. Is D a fixed set of t?

9.3.1 A structure for the set of all transformations on a set

The objective of this section and Section 9.3.2 is to show that the collection (set) of all transformations on every set forms a group with respect to an appropriate operation. In this section, we will illustrate the ideas on a specific set with three elements and in Section 9.3.2 we will generalize for all sets.

We consider the set $M = \{A,B,C\}$. Since we wish to consider the set of all transformations on M, the question of how many there are is relevant. This is easily answered by considering the following questions.

1. How many choices do we have as an image of A?   Answer: 3.
2. Once an image has been chosen for A, how many choices remain for images of B so that we achieve the one-to-one property?   Answer: 2.
3. Finally, once images have been chosen for A and B, how many choices remain for images of C so that we maintain the one-to-one property?   Answer: 1.
4. In how many ways can we choose images from $\{A,B,C\}$ for A, B, and C so that the mappings defined are one-to-one and onto (that is, transformations of M)?   Answer: $6 = 3 \cdot 2 \cdot 1$.

It is easy to generalize from this and conclude that the number of transformations on a finite set Q with n elements is $n(n-1)(n-2)$ $\cdots 3 \cdot 2 \cdot 1$.

The six transformations on M are suggested in Table 9.1.

TABLE 9.1   The set of transformations on $\{A,B,C\}$

Transformation	Domain element		
	A	B	C
t_1	A	B	C
t_2	A	C	B
t_3	B	A	C
t_4	B	C	A
t_5	C	A	B
t_6	C	B	A

We will consider t_2 to illustrate how to read Table 9.1: t_2 is the function from M onto M (transformation of M) defined by

$$t_2 = \{(A,A),(B,C),(C,B)\}.$$

That is, t_2 is the function which can be diagrammed as

$$
\begin{array}{ccc}
A & B & C \\
\downarrow & \downarrow & \downarrow \\
A & C & B
\end{array}
$$

An abbreviated notation which we will find convenient defines t_2 by $\begin{pmatrix} ABC \\ ACB \end{pmatrix}$.   The reader should note that all three of the notations for t_2 indicate that $t_2(A) = A$, $t_2(B) = C$, and $t_2(C) = B$. Also from Table 9.1 we determine that

$$t_3 = \{(A,B),(B,A),(C,C)\},$$

or diagrammatically,

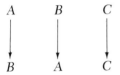

or in the abbreviated notation $t_3 = \begin{pmatrix} ABC \\ BAC \end{pmatrix}$. Note that $t_3(A) = B$, $t_3(B) = ?$, and $t_3(C) = ?$. The reader can now determine from the table that T_M, the set of all transformations on M, is

$$T_M = \left\{ \begin{pmatrix} ABC \\ ABC \end{pmatrix}, \begin{pmatrix} ABC \\ ACB \end{pmatrix}, \begin{pmatrix} ABC \\ BAC \end{pmatrix}, \begin{pmatrix} ABC \\ BCA \end{pmatrix}, \begin{pmatrix} ABC \\ CAB \end{pmatrix}, \begin{pmatrix} ABC \\ CBA \end{pmatrix} \right\}$$

$$= \{t_1, t_2, t_3, t_4, t_5, t_6\}.$$

We next define a method of combining two transformations of M (for all sets M) in Definition 9.8.

Definition 9.8 For every set M and for every f, $g \in T_M$, the *composite* of f and g (denoted by $f \circ g$) is the transformation which results from first transforming M by g followed by transforming M by f. For every $x \in M$, the image of x under $f \circ g$ is denoted by $f \circ g(x)$ or $f(g(x))$.

(*Note:* Although we omit the proof, $\circ$ is a binary operation on T_M for every set M. We refer to this binary operation as *composition* of transformations. For every f, $g \in T_M$, we read $f \circ g$ as "f composition g" or as "the composite of f and g.")

Some illustrations which demonstrate a procedure for determining the composite of transformations are shown in Example 9.7.

EXAMPLE 9.7 Determine $t_4 \circ t_3$; that is, $\begin{pmatrix} ABC \\ BCA \end{pmatrix} \circ \begin{pmatrix} ABC \\ BAC \end{pmatrix}$. We must determine the images of A, B, and C under the transformation of M which results from first transforming M by t_3 followed by transforming M by t_4.

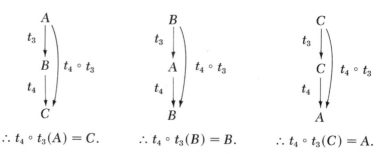

$$\therefore t_4 \circ t_3(A) = C. \qquad \therefore t_4 \circ t_3(B) = B. \qquad \therefore t_4 \circ t_3(C) = A.$$

Therefore, $t_4 \circ t_3$ has the mapping diagram:

A   B   C
↓   ↓   ↓
C   B   A

which in our abbreviated notation is denoted by $\begin{pmatrix} ABC \\ CBA \end{pmatrix}$. That is, $t_4 \circ t_3 = t_6$.

The use of mapping diagrams for t_4 and t_3 simplifies the computation of $t_4 \circ t_3$ as follows:

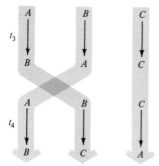

The shaded arrows indicate that $t_4 \circ t_3$ has the following mapping diagram:

A   B   C
↓   ↓   ↓
C   B   A

Therefore, we can conclude that $t_4 \circ t_3 = \begin{pmatrix} ABC \\ CBA \end{pmatrix}$. Or, we can simplify the problem further by capitalizing on our notation as follows:

$$t_3 = \begin{pmatrix} A & B & C \\ B & A & C \end{pmatrix}$$

$$t_4 = \begin{pmatrix} A & B & C \\ B & C & A \end{pmatrix}$$

The arrows clearly indicate that $t_4 \circ t_3 = \begin{pmatrix} ABC \\ CBA \end{pmatrix}$.

We give a further illustration in Example 9.8.

EXAMPLE 9.8

a. Determine $\begin{pmatrix} ABC \\ CBA \end{pmatrix} \circ \begin{pmatrix} ABC \\ ACB \end{pmatrix}$.

Solution:

$$\begin{pmatrix} A & B & C \\ A & C & B \end{pmatrix}$$

$$\begin{pmatrix} A & B & C \\ C & B & A \end{pmatrix}$$

Therefore, $\begin{pmatrix} ABC \\ CBA \end{pmatrix} \circ \begin{pmatrix} ABC \\ ACB \end{pmatrix} = \begin{pmatrix} ABC \\ CAB \end{pmatrix}$.

Exercise set 9.4

1. a. Write the operation table for $\circ$ on
$$T_M = \left\{ \begin{pmatrix} ABC \\ ABC \end{pmatrix}, \begin{pmatrix} ABC \\ ACB \end{pmatrix}, \begin{pmatrix} ABC \\ BAC \end{pmatrix}, \begin{pmatrix} ABC \\ BCA \end{pmatrix}, \begin{pmatrix} ABC \\ CAB \end{pmatrix}, \begin{pmatrix} ABC \\ CBA \end{pmatrix} \right\}.$$

   b. Is there an identity with respect to $\circ$ in T_M? Justify your answer.

   c. Does every element of T_M have an inverse with respect to $\circ$? Justify your answer.

   d. Is $\circ$ commutative? Explain your answer.

   e. Determine the following:

   (1) $\left[\begin{pmatrix} ABC \\ ACB \end{pmatrix} \circ \begin{pmatrix} ABC \\ CBA \end{pmatrix} \right] \circ \begin{pmatrix} ABC \\ BAC \end{pmatrix}$

   (2) $\begin{pmatrix} ABC \\ ACB \end{pmatrix} \circ \left[\begin{pmatrix} ABC \\ CBA \end{pmatrix} \circ \begin{pmatrix} ABC \\ BAC \end{pmatrix} \right]$

   (3) Can you determine from (1) and (2) whether $\circ$ is associative?

    f. How many cases such as (1) and (2) in part e would you have to investigate in order to determine whether ∘ is associative?

    g. Assuming that ∘ is associative, is T_M, ∘ a group? Explain.

    h. Is T_M, ∘ a commutative group? Explain.

2. Let $N = \{A,B,C,D\}$.

    a. How many transformations are there on N?

    b. List all of the transformations T_N on N using the notational scheme as in Exercise 1.

    Example. One transformation on N is $\begin{pmatrix} ABCD \\ ABDC \end{pmatrix}$ which indicates that A is mapped to A, B to B, C to D, and D to C.

    c. Determine the operation table for ∘ on T_N.

    d. What is the identity with respect to ∘?

    e. Does every element of T_N have an inverse with respect to ∘? Determine the inverse of each.

    f. Is ∘ commutative? Explain.

9.3.2 Structural properties of ∘

The examples and exercises of the previous section suggest that the set of transformations on a given set has the following properties with respect to the operation composition of transformations: There is an identity, every element has an inverse, and the operation is not commutative. This is true of the set of transformations on every set except for a two element set.

Definition 9.9 For every set S, the transformation i on S, which maps every element to itself, is called the *identity transformation* on S.

    For every transformation t on a set S, t is one-to-one and onto; for this reason it is true that for every $y \in S$, there *exists a unique* $x \in S$ such that $t(x) = y$. That is, every element y of S has a unique element x of S mapped to it. For this reason there *exists* a *unique* transformation, call it t^{-1}, which is related to t as follows:

For every $y \in S$, if $t(x) = y$, then $t^{-1}(y) = x$.

That is, t^{-1} maps every element of S "back to where it came from."

We illustrate for $S = \{A,B,C,D\}$ and t and t^{-1} defined by the accompanying mapping diagrams.   From the diagrams we can deter-

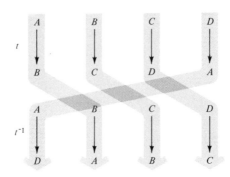

mine the following:

$t(A) = B$, $t^{-1}(B) = A$, and $t^{-1} \circ (A) = A$;

$t(B) = C$, $t^{-1}(C) = B$, and $t^{-1} \circ t(B) = B$;

$t(C) = D$, $t^{-1}(D) = C$, and $t^{-1} \circ t(C) = C$;

$t(D) = A$, $t^{-1}(A) = D$, and $t^{-1} \circ (D) = D$.

That is, $t^{-1} \circ t = i$.   We leave to the reader the problem of showing that $t \circ t^{-1} = i$.

Definition 9.10   For every set S, for every transformation t on S, the *inverse of t* (with respect to $\circ$), denoted by t^{-1}, is the transformation on S such that for every x, $y \in S$, $t^{-1}(y) = x$ if and only if $t(x) = y$; that is, $t \circ t^{-1} = i$ and $t^{-1} \circ t = i$.

It remains to investigate more carefully whether $\circ$, composition of transformations, is an associative operation.   Suppose t_1, t_2, and t_3 are transformations of a set S.   We must determine whether for every $x \in S$, the image of x, as a result of transforming S by $(t_1 \circ t_2) \circ t_3$ is the same as when S is transformed by $t_1 \circ (t_2 \circ t_3)$.
For every $x \in S$, there exists $y \in S$ such that $t_3(x) = y$.   That is,

$$t_3 = \begin{pmatrix} \cdot & \cdot & \cdot & x & \cdot & \cdot & \cdot \\ \cdot & \cdot & \cdot & y & \cdot & \cdot & \cdot \end{pmatrix}.$$

Similarly, there exists $z \in S$ such that

$$t_2 = \begin{pmatrix} \cdot & \cdot & \cdot & y & \cdot & \cdot & \cdot \\ \cdot & \cdot & \cdot & z & \cdot & \cdot & \cdot \end{pmatrix}.$$

Therefore,

$$t_2 \circ t_3 = \begin{pmatrix} \cdot & \cdot & \cdot & x & \cdot & \cdot & \cdot \\ \cdot & \cdot & \cdot & z & \cdot & \cdot & \cdot \end{pmatrix}.$$

Moreover, there exists $w \in S$, such that

$$t_1 = \begin{pmatrix} \cdot & \cdot & \cdot & z & \cdot & \cdot & \cdot \\ \cdot & \cdot & \cdot & w & \cdot & \cdot & \cdot \end{pmatrix}.$$

So that,

$$t_1 \circ (t_2 \circ t_3) = \begin{pmatrix} \cdot & \cdot & \cdot & x & \cdot & \cdot & \cdot \\ \cdot & \cdot & \cdot & w & \cdot & \cdot & \cdot \end{pmatrix}.$$

Using this notation, we consider $(t_1 \circ t_2) \circ t_3$.

$$t_3 = \begin{pmatrix} \cdot & \cdot & \cdot & x & \cdot & \cdot & \cdot \\ \cdot & \cdot & \cdot & y & \cdot & \cdot & \cdot \end{pmatrix}$$

$$t_2 = \begin{pmatrix} \cdot & \cdot & \cdot & y & \cdot & \cdot & \cdot \\ \cdot & \cdot & \cdot & z & \cdot & \cdot & \cdot \end{pmatrix}$$

$$t_1 = \begin{pmatrix} \cdot & \cdot & \cdot & z & \cdot & \cdot & \cdot \\ \cdot & \cdot & \cdot & w & \cdot & \cdot & \cdot \end{pmatrix}$$

$$\therefore t_1 \circ t_2 = \begin{pmatrix} \cdot & \cdot & \cdot & y & \cdot & \cdot & \cdot \\ \cdot & \cdot & \cdot & w & \cdot & \cdot & \cdot \end{pmatrix}.$$

$$\therefore (t_1 \circ t_2) \circ t_3 = \begin{pmatrix} \cdot & \cdot & \cdot & x & \cdot & \cdot & \cdot \\ \cdot & \cdot & \cdot & w & \cdot & \cdot & \cdot \end{pmatrix}.$$

Although the illustration above does not constitute a proof, it strongly suggests that:

For every $x \in S$, $(t_1 \circ t_2) \circ t_3(x) = t_1 \circ (t_2 \circ t_3)(x)$.

That is, apparently composition of transformations is an associative operation. The formal statement and proof is as follows:

For every set S, for every $x \in S$, and for all $t_1, t_2, t_3 \in T_S$, $(t_1 \circ t_2)$ $\circ t_3(x) = t_1 \circ (t_2 \circ t_3)(x)$.

PROOF

1. $x \in S, t_1, t_2, t_3 \in T_S$.	1. Given
2. $(t_1 \circ t_2) \circ t_3(x) = t_1 \circ t_2(t_3(x))$.	2. Definition of $\circ$

3.	$= t_1(t_2(t_3(x)))$.	3.	Why?
4.	$\therefore (t_1 \circ t_2) \circ t_3(x) = t_1(t_2(t_3(x)))$.	4.	?
5.	$t_1 \circ (t_2 \circ t_3)(x) = t_1(t_2 \circ t_3(x))$.	5.	?
6.	$= t_1(t_2(t_3(x)))$.	6.	?
7.	$\therefore t_1 \circ (t_2 \circ t_3)(x) = t_1(t_2(t_3(x)))$.	7.	?
8.	$\therefore (t_1 \circ t_2) \circ t_3(x) = t_1 \circ (t_2 \circ t_3)(x)$.	8.	?

Therefore, the set of all transformations on a set S forms a group with respect to the operation of composition of transformations.

Definition 9.11   Let S be a set with n elements, where n is a non-zero whole number.   The group of transformations on S is denoted by S_n, $\circ$ and is called the *transformation group on S*, the *permutation group on S*, or the *symmetric group on n elements*.

We leave to the reader the problem of answering the question of whether S_n, $\circ$ is a commutative or noncommutative group.   If S is any set, we denote the transformation group on S by T_S, $\circ$.

Exercise set 9.5

1. Use group properties to prove that for every set S, for every $t_1, t_2 \in T_S$:

   a. $t_1^{-1} \circ (t_1 \circ t_2) = t_2$.
   b. $(t_1 \circ t_2) \circ t_2^{-1} = t_1$.
   c. $(t_1 \circ t_2)^{-1} = t_2^{-1} \circ t_1^{-1}$.

transformations of a plane— motion geometry

10.1 INTRODUCTION AND GENERAL REMARKS

Euclid's famous treatise, "The Elements," in which he presented to the mathematical world the first attempt to place the study of geometry on a sound deductive basis, indicates that Euclid thought of geometric figures as being movable. He suggested that two triangles are congruent if one can be superimposed upon the other. We emphasized in our previous work with geometry that geometric figures are sets of points which in turn are abstract ideas with postulated properties; therefore, they are not things that can be moved about.

On the other hand, the terminology we use in connection with functions also is suggestive of motion. We refer to domain elements of a function as being mapped to elements of the range. This suggests the possibility of using the concept of function to describe what we mean by a movement of a point or a figure in a plane.

Our previous work with congruence suggests that the models of triangles in Fig. 10.1 are models of congruent triangles. Since these are, in fact, models of triangles, they enjoy the property of being movable. To convince ourselves that the two models have

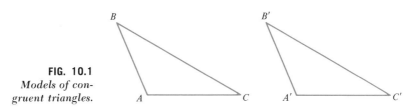

FIG. 10.1
Models of con-
gruent triangles.

the same size and shape (that is, are models of congruent triangles), it is natural to think of placing one model on the other to see if "they fit."

Let us investigate how the moving of $\triangle ABC$ onto $\triangle A'B'C'$ can be described mathematically by abstracting from the physical situation. If the models of $\triangle ABC$ and $\triangle A'B'C'$ do in fact "fit," then every model of a point of the model of $\triangle ABC$ falls on a model of a point of the model of $\triangle A'B'C'$. In a mathematical sense what we have done is to map the set of points which is $\triangle ABC$ one-to-one onto the set of points which is $\triangle A'B'C'$. That is, mathematically we think of mapping the points of $\triangle ABC$ onto the points of $\triangle A'B'C'$ rather than moving $\triangle ABC$ and placing it on top of $\triangle A'B'C'$.

Now the question arises as to whether it is natural to think of mapping the points of a triangle in a plane α onto the points of another triangle in α independently of mapping *all* of the points of α one-to-one onto points of α. This would constitute sort of a "ripping" of the triangle out of the plane and placing it at another location. Probably it is more accurate to think in terms of a one-to-one mapping of the plane *onto* the plane which in turn determines the one-to-one mapping of $\triangle ABC$ onto $\triangle A'B'C'$. What we are talking about is, of course, a transformation of the plane.

In Fig. 10.1 the model of $\triangle ABC$ could be moved to fit on the model of $\triangle A'B'C'$ by sliding $\triangle ABC$ along a line. Is the same true about the triangles in Fig. 10.2? (See p. 238.)

The considerations related to Figs. 10.1 and 10.2 are suggestive of activities that are useful for elementary school students. To investigate properties of geometric figures, teachers and students can use models which they slide, flip, turn, and so on. The question of exactly how many different kinds of motions there are which do not change the size or shape of a geometric figure arises. In the next sections we turn our attention to a more precise description of the concept of motion in geometry and to the answer of the question above.

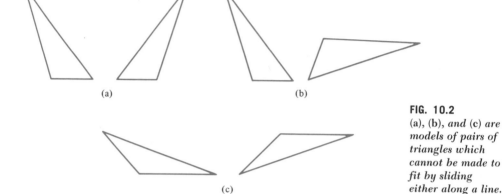

(a)   (b)

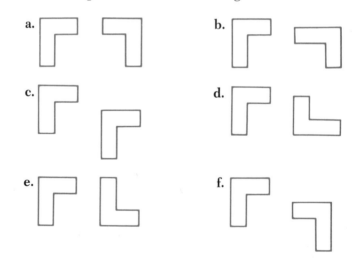

(c)

FIG. 10.2
(a), (b), *and* (c) *are
models of pairs of
triangles which
cannot be made to
fit by sliding
either along a line.*

Exercise set 10.1

1. Describe the motion necessary to move the model on the left to fit on top of the model on the right.

a.

b.

c.

d.

e.

f.

10.2 RIGID MOTIONS OF A PLANE

We will explore further the idea of motions, or more accurately transformations, of a plane. More specifically, we wish to define precisely the notion of transformations of a plane which correspond to the various movements of models of geometric figures which do not change their size and shape. If the distance between every two points of a geometric figure is preserved, then its "size and shape" will also be preserved. What we are interested in is the set of all transformations of a plane which preserve the distance between every two points. Such transformations of a plane are called *rigid motions*.

Definition 10.1 A transformation t of a plane α is a *rigid motion* of α if and only if for every two points $A, B \in \alpha$, $AB = t(A)t(B)$.

Theorem 10.1 For every rigid motion t of a plane α and for all points $A, B, C \in \alpha$, if A–B–C, then $t(A)$–$t(B)$–$t(C)$.

PROOF

1. t is a transformation of α, A, B, $C \in \alpha$, and A–B–C.
2. $\therefore AB + BC = AC$.
3. $AB = t(A)t(B)$, $BC = t(B)t(C)$, and $AC = t(A)t(C)$.
4. $\therefore t(A)t(B) + t(B)t(C) = t(A)t(C)$.
5. $\therefore t(A)$–$t(B)$–$t(C)$.

1. Why?

2. ?
3. ?

4. ?
5. ?

Corollary 10.1.1 For every rigid motion t of a plane α and for every line segment AB in α, the set of image points of $\overline{AB}$ is a line segment $A'B'$ and $\overline{AB} \cong \overline{A'B'}$.

Corollary 10.1.2 For every rigid motion t of a plane α and for every ray AB in α, the set of image points of $\overrightarrow{AB}$ is a ray.

The proofs of the corollaries are left to the reader.

We emphasize the real importance of Theorem 10.1 and its corollaries: By definition of a rigid motion, the "size" of geometric figures is left unchanged under a rigid motion, and Theorem 10.1 and its corollaries guarantee that the "shape" of figures is also unchanged under rigid motions.

Hereafter, in this chapter, if t is a transformation of a plane α, then for every point $A \in \alpha$, we will denote the image of A by A'.

10.3   TRANSLATIONS OF A PLANE

Our reference to Fig. 10.1 asked the reader to imagine sliding a model triangle along a line to test its fit with another triangle. Our purpose in this section is to give mathematical precision to the idea of "sliding points along a line." Since points are geometric abstractions, not endowed with the property of being movable, we think of this "sliding movement" in terms of a transformation. The transformation which describes the "sliding movement" is called a *translation;* a translation maps every point in the plane to a point in the plane such that: (1) the points and their images determine rays which are parallel to a given line, (2) the measure of the distance between every point and its image is the same, and (3) the "direction" from every point to its image is the same. In the definition of translation we will use the concept of the direction of a ray. We take this—the *direction of a ray*—to be an undefined concept for our development.

Definition 10.2   A transformation t of a plane α is a *translation* of α with *directrix* line $\mathscr{L}$ (or along line $\mathscr{L}$) if and only if:

1.   For every point $B \in \alpha$, $\overleftrightarrow{BB'} \parallel \mathscr{L}$.
2.   There exists $r \in \mathbb{R}$, $r \geqslant 0$, such that for every point $B \in \alpha$, $BB' = r$.
3.   For all points $B, C \in \alpha$, $\overrightarrow{BB'}$ and $\overrightarrow{CC'}$ have the same direction.

The real number r is the *distance* of the translation and for every point $B \in \alpha$, the direction of $\overrightarrow{BB'}$ is the *direction* of the translation.

   Let us discuss Definition 10.2 in intuitive language. Property 1 says that every point B in the plane is "slid" to a point B' in the plane such that the line determined by points B and B' is parallel to $\mathscr{L}$. Property 2 guarantees that every point in the plane is "slid" the same distance and Property 3 guarantees that every point is "slid" in the same direction.

   We illustrate the properties of Definition 10.2 in Example 10.1.

EXAMPLE 10.1   The accompanying model suggests the following:

$\overleftrightarrow{BB'} \parallel \mathscr{L}, \quad \overleftrightarrow{CC'} \parallel \mathscr{L}, \quad \overleftrightarrow{DD'} \parallel \mathscr{L};$
$BB' = CC' = DD';$
$\overrightarrow{BB'}, \overrightarrow{CC'},$ and $\overrightarrow{DD'}$ all have the same direction.

Now since our goal in this chapter is to describe all possible rigid motions of a plane, we consider the question of whether every translation is a rigid motion in Theorem 10.2.

Theorem 10.2 For every translation t of a plane α, t is a rigid motion of α. That is, for all points $B, C \in \alpha$, $BC = B'C'$. (Reference to the accompanying model will help in following the proof.)

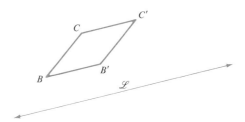

PROOF

1.	t is a translation of α.	1.	Given
2.	$\therefore$ For all points $B, C \in \alpha$, $BB' = CC'$.	2.	Definition of translation
3.	$\therefore m(\overline{BB'}) = m(\overline{CC'})$.	3.	Definition of measure of line segments
4.	$\therefore \overline{BB'} \cong \overline{CC'}$.	4.	Definition of $\cong$ for line segments
5.	$\overleftrightarrow{BB'} \parallel \mathscr{L}$ and $\overleftrightarrow{CC'} \parallel \mathscr{L}$.	5.	Definition of translation
6.	$\therefore \overleftrightarrow{BB'} \parallel \overleftrightarrow{CC'}$.	6.	Symmetric, transitive properties of $\parallel$
7.	$\therefore CBB'C'$ is a parallelogram.	7.	Why?
8.	$\therefore \overline{BC} \cong \overline{B'C'}$.	8.	Exercise 5 of Exercise Set 6.5, p. 154
9.	$\therefore m(\overline{BC}) = m(\overline{B'C'})$.	9.	?
10.	$\therefore BC = B'C'$.	10.	?

As a result of Exercise Set 10.1 the reader is probably convinced that there is more than one kind of rigid motion; that is, translations are not the only transformations of the plane which preserve distance. Since we will be studying more than one kind of rigid motion, it will be necessary to have a criterion for determining when a transformation is a translation. This is given in Theorem 10.3.

Theorem 10.3 A transformation t of a plane α, is a translation of α if and only if every line segment CD in α is mapped by t to a line segment $C'D'$ such that $\overline{CD} \cong \overline{C'D'}$ and $\overline{CD} \parallel \overline{C'D'}$.

The proof uses many of the same ideas as those in the proof of Theorem 10.2 and is left as an exercise for the reader.

Another way that we can recognize whether a transformation of a plane is a translation is by investigating its fixed points and fixed sets. Since our definition of translation permits the measure of the distance r between every point and its image to be zero, we consider the identity transformation to be a translation; the identity leaves every point fixed. However, it can be proved that if a translation moves at least one point of the plane, then it moves every point of the plane. The *fixed sets* of a translation are all the lines which are parallel to the directrix of the transformation. Therefore, every translation, other than the identity, has no fixed points but has infinitely many lines as fixed sets.

10.3.1 Composition of translations

Since translations of a plane are special kinds of transformations, composition of translations of a plane is defined. We will determine whether the set of all translations of a plane together with composition forms a group. We began by investigating the question of whether the composition of two translations is again a translation; that is, the question of whether the set of translations is closed with respect to composition. First, suppose that t_1 and t_2 are two translations along the same line $\mathscr{L}$ as suggested in Fig. 10.3. Since t_2 is a translation, it follows that for all points $B, C \in \alpha$ $\overline{BC} \cong \overline{B'C'}$ and $\overline{BC} \parallel \overline{B'C'}$; since t_1 is a translation, it follows that $\overline{B'C'} \cong \overline{B''C''}$ and $\overline{B'C'} \parallel \overline{B''C''}$. Therefore, $\overline{BC} \cong \overline{B''C''}$ and $\overline{BC} \parallel \overline{B''C''}$. Why? Therefore, $t_1 \circ t_2$ is a translation of α. Next

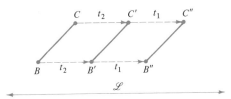

suppose that t_1 and t_2 are translations along different lines $\mathscr{L}_1$ and $\mathscr{L}_2$ as suggested in Fig. 10.4.   Since t_2 and t_1 are translations, the following are true:

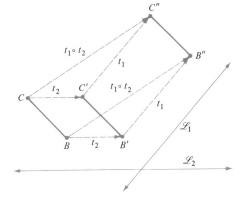

$\overline{BC} \cong \overline{B'C'}$ and $\overline{B'C'} \cong \overline{B''C''}$.
$\therefore \overline{BC} \cong \overline{B''C''}$.
$\overline{BC} \parallel \overline{B'C'}$ and $\overline{B'C'} \parallel \overline{B''C''}$.
$\therefore \overline{BC} \parallel \overline{B''C''}$.
$\therefore t_1 \circ t_2$ is a translation of α.

The above discussion proves that the set of all translations of a plane α is closed with respect to composition of transformations. Moreover, since composition of transformations in general is well defined, composition of translations is well defined.   Composition of translations is associative since composition is an associative operation on the set of all transformations of a plane.   We also have the identity translation—the translation which leaves every point fixed.   Finally, we consider the question of whether every translation has an inverse.   Clearly, if t is a translation of a plane α along a line $\mathscr{L}$ which maps every point A to a point A' such that $AA' = r$, then the inverse of t is the translation t^{-1} along $\mathscr{L}$ which maps every point B to a point B' such that $BB' = r$ and

t^{-1} has the opposite direction of t.   Therefore, the set of all translations of a plane α is a group with respect to the operation of composition.   We leave to the reader the problem of determining whether this is a commutative or noncommutative group.

10.3.2   Invariants under translations

In Theorem 10.2, we proved that every translation is a rigid motion; from this it follows that every line segment in a plane is mapped by every translation of the plane to a congruent line segment.   That is, translations preserve distance between points, betweenness of points, and the measure of line segments.   A property that is preserved by a transformation is said to be *invariant* with respect to (or under) the transformation.   Our concern is with properties of sets of coplanar points which are invariant under rigid motions; in this section, our concern is specifically with some properties of coplanar points which are invariant under translations.   Do you guess that the measures of angles are invariant under translations?   What about noncollinearity of points, parallelism, and perpendicularity of lines?   The reader is asked to answer these questions in Exercise Set 10.2.

Proposition 10.1   For every translation t on a plane α and for all points A, B, $C \in \alpha$, if A, B, and C are noncollinear, then A', B', and C' are noncollinear.

The proof is left to the reader (Exercise 5 of Exercise Set 10.2).

Theorem 10.4   For every translation t on a plane α and for every $\angle ABC$ in α, the set of image points of $\angle ABC$ is an angle which is congruent to $\angle ABC$.

The Theorem requires that we prove two things:

1.   The set of image points of an angle under a translation is an angle; and
2.   The set of image points is congruent to the given angle.

     (Reference to the accompanying model will help in following the proof.)

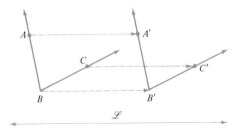

PROOF

1.	t is a translation on α and $\angle ABC$ is in α.	1.	Why?
2.	A, B, and C determine $\angle ABC$ and therefore, A, B, and C are noncollinear.	2.	?
3.	$\therefore A'$, B', and C' are noncollinear.	3.	?
4.	$\therefore A$, B, and C determine $\triangle ABC$, and A', B', and C' determine $\angle A'B'C'$ and $\triangle A'B'C'$.	4.	?
5.	$\overline{AB} \cong \overline{A'B'}$, $\overline{AC} \cong \overline{A'C'}$, and $\overline{BC} \cong \overline{B'C'}$.	5.	?
6.	$\therefore \triangle ABC \cong \triangle A'B'C'$.	6.	?
7.	$\therefore \angle ABC \cong \angle A'B'C'$.	7.	?

Exercise set 10.2

1. Let t be a translation of a plane α and A, B, and C be points in α such that A–B–C. Prove that A'–B'–C'.

2. Let t be a translation of a plane α and $\overline{AB}$ and $\overline{CD}$ be line segments in α such that $\overline{AB} \cap \overline{CD} = \{Q\}$. Prove that $\overline{A'B'} \cap \overline{C'D'} = \{Q'\}$.

3. Let t be a translation of a plane α and $\overline{AB}$ and $\overline{CD}$ be line segments in α such that $\overline{AB} \parallel \overline{CD}$. Prove that $\overline{A'B'} \parallel \overline{C'D'}$.

4. Let t be a translation of a plane α and $\overline{AB}$ and $\overline{CD}$ be line segments in α such that $\overline{AB} \perp \overline{CD}$. Prove that $A'B' \perp \overline{C'D'}$.

5. Prove Proposition 10.1.

6. Prove: If a translation t of a plane α moves at least one point of a plane, then it moves every point of the plane.

7. Draw models which suggest the image of $\triangle ABC$ (see the model on p. 246) under the composition $t_1 \circ t_2$ and $t_2 \circ t_1$, where t_1 and t_2 are translations described, respectively, as follows.

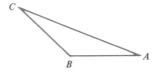

a. t_1: horizontally, to the right, 2 inches
   t_2: vertically, up, 1 inch
b. t_1: horizontally, to the right, 2 inches
   t_2: horizontally, to the right, 1 inch
c. t_1: horizontally, to the left, 2 inches
   t_2: horizontally, to the right, 1 inch
d. t_1: obliquely on the "45-degree line" up and to the right, 2 inches
   t_2: vertically, down, 1 inch
e. t_1: obliquely on the "45-degree line" up and left, 2 inches
   t_2: horizontally, to the left, 1 inch.

8. From Exercise 7, what would you *guess* about the commutativity of composition of translations of a plane?   That is, would you guess that the group of translations on a plane α is a commutative or noncommutative group?

9. Show by completing the accompanying diagram that $t_1 \circ (t_2 \circ t_3) = (t_1 \circ t_2) \circ t_3$, where t_1, t_2, and t_3 are translations of α along lines $\mathscr{L}_1$, $\mathscr{L}_2$, and $\mathscr{L}_3$, respectively.

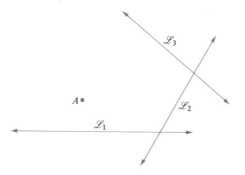

10.4   ROTATIONS OF A PLANE

The reader probably discovered in Fig. 10.2 and Exercise Set 10.1 that it is not always possible to move one of two models of congruent geometric figures to fit the other with a slide motion. In geometric terms, the reader has discovered that there are rigid motions of the plane other than translations.

In this section we consider the idea of a rotation of a plane. To do this we need some elementary notions about circles. Two coplanar circles are called *concentric circles* if and only if they have the same center. We leave as an exercise for the reader the proof of the following: If $\mathscr{C}_1$ and $\mathscr{C}_2$ are concentric circles with center P and radii r_1 and r_2, $r_2 > r_1$, respectively (see Fig. 10.5), and $\overrightarrow{PX}$ is a ray which intersects $\mathscr{C}_1$ and $\mathscr{C}_2$ at A and B, respectively, then $AB = r_2 - r_1$.

FIG. 10.5
Concentric circles
$\mathscr{C}_1$ *and* $\mathscr{C}_2$.

The models in Fig. 10.6 suggest how a geometric figure is related to its image set under a rotation about a point P.

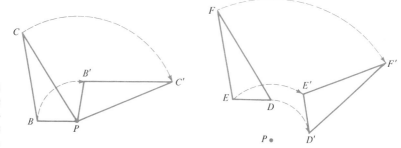

FIG. 10.6
Models to suggest the image of a triangle under a rotation.

Intuitively, a rotation of a plane is a transformation of the plane which leaves one point P of the plane fixed (the center of the rotation) and maps every other point X in the plane to a point X' such that X and X' are on the same circle with center at P. Also every point in the plane, except the center, is rotated in the same "direction" and by the same "amount." We consider the term "*direction*" of a rotation as an *undefined concept*. We will refer to rotations as having *clockwise* or *counterclockwise* direction. To define the "amount" of a rotation, or more specifically the measure of a rotation, we use the concept of an arc of a circle and its measure.

Definition 10.3   Let $\mathscr{C}$ be a circle with center P, and let A and B be points on $\mathscr{C}$ which are not endpoints of a line segment through P; then:

1. The *minor arc* of $\mathscr{C}$ determined by A and B, denoted by $\overset{\frown}{AB}$, is the set consisting of points A, B, and all points on $\mathscr{C}$ in the interior of $\angle APB$.
2. The *major arc* of $\mathscr{C}$ determined by A and B, denoted by $\overset{\frown}{A\text{–}B}$, is the set consisting of points A, B, and all points on $\mathscr{C}$ in the exterior of $\angle APB$.

If A and B are endpoints of a line segment through P, then each arc of $\mathscr{C}$ determined by points A and B is called a *semicircle*.

Definition 10.4   Let $\mathscr{C}$ be a circle with center P.   Let A and B be points on $\mathscr{C}$ which are not the endpoints of a line segment through the center; then the *degree measure* of $\overset{\frown}{AB}$

$$m_\circ(\overset{\frown}{AB}) = m_\circ(\angle APB) \text{ if } A \neq B, \ m_\circ(\overset{\frown}{AB}) = 0 \text{ if } A = B$$

and

$$m_\circ(\overset{\frown}{A\text{–}B}) = {}^+360 - m_\circ(\angle APB).$$

If A and B are endpoints of a line segment through the center, then

$$m_\circ(\overset{\frown}{AB}) = {}^+180.$$

Finally, $m_\circ(\mathscr{C}) = {}^+360$.

EXAMPLE 10.2   Let A and B be points on a circle as sug-

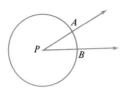

gested in the accompanying model.   Let $m_\circ(\angle APB) = {}^+34$.   Then, $m_\circ(\overset{\frown}{AB}) = m_\circ(\angle APB) = {}^+34$, and $m_\circ(\overset{\frown}{A\text{–}B}) = {}^+360 - {}^+34$.

We will usually not be concerned whether an arc is a major arc or a minor arc.   Since Definition 10.4 defines the measure of every arc of a circle, when we refer to an arc and its measure we will denote them by $\widehat{AB}$ and $m_\circ(\widehat{AB})$, respectively.   Moreover, $m_\circ(\widehat{AB})$ is a real number r such that $0 \leqslant r \leqslant {}^+360$.

Definition 10.5   A transformation t of a plane α is a *rotation* of α with *center P* (or around point P) if and only if:

1.   For every point $B \in \alpha$, B and B' are points on the same circle with center P.
2.   There exists $r \in \mathbb{R}$, $0 \leqslant r \leqslant {}^+360$, such that for every point $B \in \alpha$, $m_\circ(\widehat{BB'}) = r$.

The real number r is called the *degree measure* (or simply, *measure*) of the rotation.

   Now we consider the question of whether every rotation of a plane α is a rigid motion of the plane.

Theorem 10.5   For every point P in a plane α and for every rotation t of a plane α around P, t is a rigid motion of α.   That is, for all points $A, B \in \alpha$, $AB = A'B'$.   (Reference to the accompanying model will help in following the proof.)

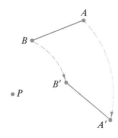

   There are three cases to consider for points A, B, and P.

1. A, B, and P are distinct and collinear,
2. A or B equals P, and
3. A, B, and P are distinct and noncollinear.

We leave the first two cases as exercises for the reader.

PROOF OF CASE 3

1. Points A, B, and P are noncollinear.	1. Assumption
2. t is a rotation of α.	2. Given

3. ∴ A and A' are points on a circle $\mathscr{C}_1$ with   3. Definition of
    center P; B and B' are points on a circle   rotation
    $\mathscr{C}_2$ with center P.
4. ∴ $PA = PA'$ and $PB = PB'$.   4. Why?
5. ∴ $\overline{PA} \cong \overline{PA'}$ and $\overline{PB} \cong \overline{PB'}$.   5. ?
6. $\angle BPA \cong \angle B'PA'$.   6. Exercise 4 of
                                                                       Exercise Set
                                                                       4.5, p. 110
7. ∴ $\triangle PBA \cong \triangle PB'A'$.   7. ?
8. ∴ $\overline{BA} \cong \overline{B'A'}$.   8. ?
9. ∴ $BA = B'A'$.   9. ?

Since every rotation of a plane α is a rigid motion, it follows from Theorem 10.5 that the image of a line segment under a rotation is a congruent line segment and that the image set of a ray under a rotation is a ray.   It is easy to prove as a result of these facts that the measure of angles is invariant under rotations.

We next turn our attention to giving a criterion for determining when a translation of a plane α is a rotation.   One difference between a translation and a rotation becomes apparent when we study their fixed points and fixed sets.   Recall that a translation, other than the identity, has no fixed points.   A rotation, other than the identity (a rotation with measure zero), has exactly one fixed point —the center P of the rotation.   It is easy to see that every circle with center P is a fixed set of every rotation with center P.   Moreover, as illustrated in Fig. 10.7, every line through point P is also a fixed set under a rotation around P with measure $^+180$.

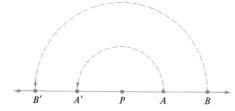

FIG. 10.7
Model to suggest that every line through point P is fixed under a rotation around P.

We emphasize: A translation has no fixed points and a rotation has one fixed point; both have infinitely many fixed sets.

Another criterion by which a rotation can be recognized is given in Theorem 10.6.   A more complete development of geometry than we have given is required for a proof; therefore, we state it without proof.

Theorem 10.6 A transformation t of a plane α, is a rotation of α with center P if and only if for all points A and B in α which are not fixed points of t, the perpendicular bisectors of $\overline{AA'}$ and $\overline{BB'}$ pass through P.

Figure 10.8 illustrates the theorem.

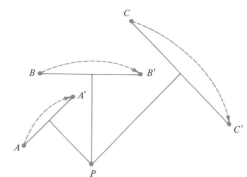

FIG. 10.8
Model to suggest
Theorem 10.8.

10.4.1 Composition of rotations

We discovered that the set of all translations of a plane forms a group under the operation of composition of transformations. Is the same true of all rotations of a plane?

First we consider the set of all rotations about *one* point in the plane. It is clear that one rotation followed by another rotation both about a point P is again a rotation about P. The rotation with center P and measure zero is the identity, and if t is a rotation in a clockwise direction with measure r, $r > 0$, then t^{-1}, the rotation around P in a clockwise direction with measure $^+360 - r$, is the inverse of t with respect to composition. That is, $t \circ t^{-1} = t^{-1} \circ t = i$, where i is the identity transformation. Moreover, we know that composition is an associative operation; therefore, the set of all rotations around a given point forms a group under the operation of composition of transformations. We leave to the reader the problem of investigating whether the group is commutative.

The situation for the composition of two rotations about different points is not so simple. Reference to Fig. 10.9 suggests that, in some cases at least, the composition of a rotation around P_1 followed by a rotation around P_2, $P_1 \neq P_2$, is again a rotation. The center of

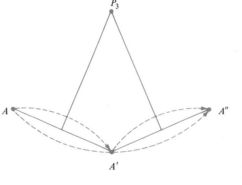

FIG. 10.9
Model to suggest a case for which the composition of two rotations about different points is a rotation.

the rotation is the point P_3 which is the point at which the perpendicular bisectors of $\overline{AA'}$ and $\overline{A'A''}$ intersect.

However, consider the composition of a rotation t_1 about P_1 with measure $^+150$ followed by a rotation t_2 about a point P_2 with measure $^+210$ as illustrated in Fig. 10.10.

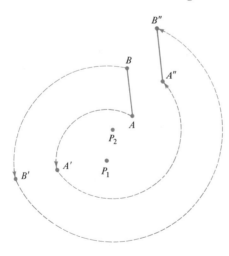

FIG. 10.10
Model to suggest a case for which the composition of two rotations about different points is not a rotation.

Apparently $\overline{AB} \cong \overline{A''B''}$ and $\overline{AB} \parallel \overline{A''B''}$ so that the composition of these two rotations is a translation. Thus, it appears that the composition of two rotations is either a rotation or a translation. Let us investigate the compositions of rotations and translations;

maybe the union of the set of translations of a plane with the set of rotations of a plane forms a group with respect to the operation of composition.

Let r and t be a rotation and a translation of a plane α, respectively. We will investigate whether the transformation of α determined by first transforming α by r and then by t is either a translation or a rotation; that is, we answer the question of whether $t \circ r$ is either a translation or a rotation. The question of whether $r \circ t$ is either a rotation or a translation we leave to the reader.

The model in Fig. 10.11 suggests a rotation r of a plane α followed by a translation t; that is, a transformation $t \circ r$ of α. The

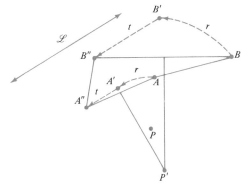

FIG. 10.11
Model to suggest
the composition of
a rotation followed
by a translation.

model suggests that points A and B are mapped to A' and B', respectively, by the rotation r with center at P; following this, A' and B' are mapped by translation t to A'' and B'', respectively, along line $\mathscr{L}$. Now in considering $t \circ r$, our concern is with points A and B, and their images under $t \circ r$, A'' and B''. The model clearly suggests that $\overline{AB}$ is not parallel to $\overline{A''B''}$; therefore, $t \circ r$ is not a translation. On the other hand, the model suggests that the perpendicular bisectors of $\overline{AA''}$ and $\overline{BB''}$ pass through a point P' for all points A and B in α. Therefore, it is apparent that $t \circ r$ is a rotation. The inferences which we made from the model are in fact provable although we omit their proofs here. Moreover, it is possible to prove that the measure of the rotation $t \circ r$ is the same as the measure of the rotation r.

Since we have established that the set of all translations and rotations of a plane α is closed with respect to the operation of composition, it is easily established that this system forms a group. Clearly the identity transformation is the identity for the system.

If $t \circ r$ is the rotation which is the transformation of α which results from first rotating the plane by r followed by translating it by t, then the inverse of $t \circ r$ is $r^{-1} \circ t^{-1}$, where t^{-1} is the inverse translation of t and r^{-1} is the inverse rotation of r. That is, the "undoing" of first rotating the plane and then translating it is accomplished by first "undoing" the translation and then "undoing" the rotation. Some experimenting with physical models will give the reader some intuition for this. Since composition of transformations in general is associative, it is associative on the set of translations and rotations. Therefore, the set of translations and rotations on a plane α forms a group with respect to the operation of composition. We leave to the reader to experiment to determine whether this group is commutative.

Exercise set 10.3

1. Let r_{30} and r_{45} denote counterclockwise rotations of a plane α with measures $^+30$ and $^+45$, respectively. Carefully draw models to suggest whether $r_{45} \circ r_{30} = r_{30} \circ r_{45}$.

2. Let t and r be a translation and a rotation, respectively, of a plane α. Carefully draw models to suggest that $r \circ t$ is either a translation or a rotation. Describe the special case for which $r \circ t$ is a translation.

3. Let r_{30} and r_{45} be two clockwise rotations of a plane about centers P_1 and P_2, $P_1 \neq P_2$, with measures $^+30$ and $^+45$, respectively. By construction determine the center of $r_{30} \circ r_{45}$.

4. Let r_{45} be a counterclockwise rotation of a plane α about a center P, let t be a translation of α. By construction, determine the center of $r_{45} \circ t$ and $t \circ r_{45}$.

5. Draw models which suggest the image of $\triangle ABC$ (see the accompanying model) under a transformation of the plane containing $\triangle ABC$ as follows.

a. A clockwise rotation around A with measure:
(1) $^+90$ (2) $^+180$ (3) $^+360$

b. A clockwise rotation around A with measure $^+90$ followed by:
  (1) a vertical translation of 1 inch
  (2) a horizontal translation of 1 inch
  (3) a rotation around C' with measure $^+180$
  (4) a rotation around P, $P \in \text{Ext}(\triangle A'B'C')$, with measure $^+90$
c. A horizontal translation of 2 inches followed by:
  (1) a counterclockwise rotation around B' with measure $^+180$
  (2) a counterclockwise rotation around A' with measure $^+90$

6. Prove: The set of image points of an angle in plane α under a rotation r is an angle congruent to the given angle.

10.5 REFLECTIONS OF A PLANE

So far we have considered translations and rotations of a plane α and have found them to be rigid motions of α. To suggest that we have not yet studied all rigid motions, we ask the reader to consider again Exercise 1a of Exercise Set 10.1, p. 238. In this exercise, it is necessary to "flip" one model in order to get it to fit the other. Our considerations in this section will be with the mathematical description of the transformations of a plane α which accomplishes such a "turning over" of geometric figures; that is, a "turning over" of the plane. Such a transformation is called a *reflection*. The word "reflection" is often used in everyday language—we speak of the reflection of trees in a clear pond, the reflection of objects in a mirror, and so on. Indeed, the following observations about an object and its reflection or image in a (flat) mirror correspond quite accurately to the defining characteristics of a reflection.

In a (flat) mirror:

(1) The image of an object *seems to be* as far behind the mirror as the object *is* in front of it.

(2) If you "look yourself in the eye" the line segment which connects each eye with its image is perpendicular to the plane of the mirror.

(3) The image of an object is exactly the same size and shape as the object.

(4) The image of an object is reversed. If you wave your *left hand*, your image in the mirror waves its *right hand*, or if you move your arm in a clockwise direction, your image moves "its arm" in a counterclockwise direction.

Some thought will convince the reader that statements (3) and (4) are deducible from (1) and (2). The mirror in the above discussion serves as a model of a plane. Since our work is with geometric figures in a plane, we will be concerned with two-sided "line mirrors." That is, we will be concerned with reflections of geometric figures in a line. The same observations as above can be made about a "line mirror" or a *line of reflection;* this is illustrated by the model in Fig. 10.12.

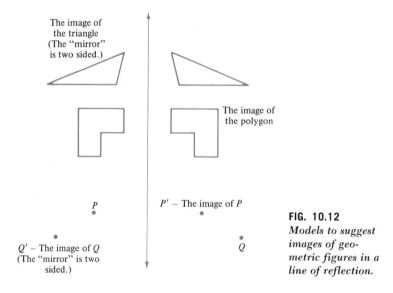

The image of
the triangle
(The "mirror"
is two sided.)

The image of
the polygon

P

P' – The image of P

Q' – The image of Q
(The "mirror" is two
sided.)

Q

FIG. 10.12
*Models to suggest
images of geo-
metric figures in a
line of reflection.*

We can now rather easily abstract from the discussion above to define a reflection of a plane α in a line $\mathcal{L}$.

Definition 10.6 A transformation of a plane α is a *reflection* of α in a line $\mathcal{L}$ if and only if for every point $A \in \alpha$, $\mathcal{L}$ is the perpendicular bisector of $\overline{AA'}$. The direction of $\overrightarrow{AA'}$ is the *direction* of the reflection, $\mathcal{L}$ is the *line of reflection.*

Now we consider the question of whether every reflection of a plane α is, as we suspect, a rigid motion of α. This question is answered in Theorem 10.7.

Theorem 10.7 Every reflection t of a plane α is a rigid motion of α. That is, for all points $B, C \in \alpha$, $BC = B'C'$. (Reference to the accompanying model will help in following the proof.)

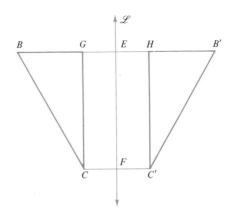

PROOF

1.	t is a reflection of α and $B, C \in \alpha$.	**1.**	Why?
2.	$\overline{BB'} \perp \mathscr{L}$ and $\overline{CC'} \perp \mathscr{L}$.	**2.**	?
3.	$\therefore \overleftrightarrow{BB'} \parallel \overleftrightarrow{CC'}$.	**3.**	?
4.	There exist perpendiculars to $\overleftrightarrow{BB'}$ from C and C', respectively. Let G and H be the points of intersection.	**4.**	?
5.	$\overline{CG} \parallel \overline{C'H}$, $\overline{CG} \parallel \mathscr{L}$, and $\overline{C'H} \parallel \mathscr{L}$.	**5.**	Exercise 11 of Exercise Set Set 4.6, p. 114
6.	$\angle CGH$ is a right angle and $\angle GHC'$ is a right angle.	**6.**	Definition of perpendicular lines
7.	$CGHC'$, $CGEF$, and $FEHC'$ are rectangles.	**7.**	?
8.	$\overline{CF} \cong \overline{GE}$, $\overline{FC'} \cong \overline{EH}$, and $\overline{GC} \cong \overline{HC'}$.	**8.**	?
9.	$\overline{BE} \cong \overline{EB'}$ and $\overline{CF} \cong \overline{FC'}$.	**9.**	?
10.	$\overline{GE} \cong \overline{EH}$.	**10.**	?

11. $\therefore \overline{BG} \cong \overline{B'H}$.	**11.**	?
12. $\therefore \angle CGB$ and $\angle C'HB$ are right angles.	**12.**	?
13. $\therefore \angle CGB \cong \angle C'HB'$.	**13.**	?
14. $\therefore \triangle BGC \cong \triangle B'HC'$.	**14.**	?
15. $\therefore \overline{BC} \cong \overline{B'C'}$.	**15.**	?
16. $\therefore BC = B'C'$.	**16.**	?

Now that we know that every reflection is a rigid motion we can, because of Theorem 10.1 (p. 239), easily determine many properties of plane figures which are invariant under reflections. We can conclude, for example, that the set of image points of a line segment under a reflection is a line segment congruent to the given line segment. It can be shown from this (we leave it as an exercise for the reader) that reflections preserve angle measure.

One criterion for the recognition of a reflection is the definition itself:

A transformation t of a plane α is a reflection of α if and only if there is a line $\mathscr{L}$ in α such that for every point $A \in \alpha$, $\mathscr{L}$ is the perpendicular bisector of $\overline{AA'}$.

A reflection, as other rigid motions, can also be characterized by its fixed points. Clearly, if t is a reflection of a plane α in a line $\mathscr{L}$, then every point of $\mathscr{L}$ is a *fixed point*. That is, a *reflection has infinitely many fixed points*. Every line which is perpendicular to the line of reflection $\mathscr{L}$ is a fixed set of the reflection; therefore, a *reflection has infinitely many fixed sets*.

10.5.1 Composition of translations, rotations, and reflections

As we did with the other rigid motions, we begin by considering the question of whether the set of all reflections of a plane forms a group with respect to the operation of composition of transformations. We first observe that the reflection which leaves every point of the plane fixed is the identity. If t is a reflection of a plane α in a line which maps every point A to a point A', then the reflection t^{-1} in $\mathscr{L}$ which maps A' back to A is the inverse of t. Since composition of transformations in general is associative, composition of reflections is associative.

Now to investigate whether the set of reflections of a plane α is closed with respect to composition, we refer to Fig. 10.13. Reference to the figure will aid the reader in proving, as suggested by

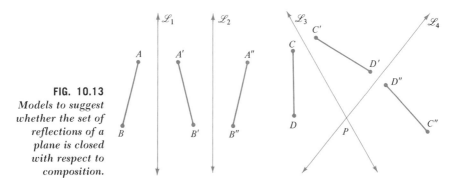

FIG. 10.13
Models to suggest whether the set of reflections of a plane is closed with respect to composition.

the models: (1) A reflection of a plane α in a line $\mathscr{L}_1$ followed by a reflection in a line $\mathscr{L}_2$, where $\mathscr{L}_1 \parallel \mathscr{L}_2$, is a translation of the plane α along a line $\mathscr{M}$ which is perpendicular to $\mathscr{L}_1$ and $\mathscr{L}_2$, (2) A reflection of α in a line $\mathscr{L}_3$ followed by a reflection of α in a line $\mathscr{L}_4$, where $\mathscr{L}_3 \cap \mathscr{L}_4 = \{P\}$, is a rotation of α with center at P. Therefore, the set of reflections of a plane α is not closed with respect to the operation of composition; and hence, the system which consists of the set of reflections of a plane α and the operation of composition of transformations does not form a group.

Now let us turn our attention to the set of all translations, rotations, and reflections of a plane α. Is this set of transformations closed with respect to composition? The question is easily answered by considering a transformation of the plane α as follows: first translate α along a line $\mathscr{L}$, and then reflect α in $\mathscr{L}$ (see Fig. 10.14).

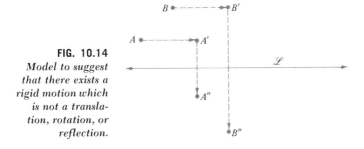

FIG. 10.14
Model to suggest that there exists a rigid motion which is not a translation, rotation, or reflection.

Our concern is with the transformation of the plane which maps A to A'' (and B to B''). We wish to determine whether it is a translation, rotation, or a reflection. We can give an intuitive answer to this question by concentrating on the fixed points and fixed sets of the transformation which, for every point A in α, maps A to A''. The translation which maps A to A' *moves* every point of the plane α, and the reflection which takes A' to A'' moves every point, except points on $\mathcal{L}$, along a line perpendicular to $\mathcal{L}$; therefore, it seems apparent that the transformation which maps A to A'' has *no fixed points*. Thus, the transformation is not a rotation. The translation along $\mathcal{L}$ has as fixed sets all lines parallel to $\mathcal{L}$ (which includes $\mathcal{L}$) but the reflection in $\mathcal{L}$ moves every point of these lines, except the points of $\mathcal{L}$, to different lines. That is, $\mathcal{L}$ is the only fixed line of the transformation which maps A to A'', translations and reflections have infinitely many lines as fixed sets; therefore, apparently the transformation is neither a translation nor a reflection.

However, for all points A and B of the plane α, $AB = A'B'$, because A' and B' are the images of A and B, respectively, under a translation; moreover, $A'B' = A''B''$ since the reflection in $\mathcal{L}$ maps A' to A'' and B' to B''; from this it follows that $AB = A''B''$; and hence, the transformation which maps A to A'' and B to B'' is a rigid motion of the plane α. Since it is the composition of a translation and a reflection, we call it a *transflection*.[1]

Definition 10.7   A transformation t of a plane α is a *transflection* of α in line $\mathcal{L}$ if and only if it is the composition of a translation along $\mathcal{L}$, followed by a reflection in $\mathcal{L}$.

A transflection is characterized by the fact that it has *no* fixed points and *one* fixed line. Another model to illustrate a transflection is given in Example 10.3.

EXAMPLE 10.3   The accompanying model illustrates the image of a triangle under a transflection.

[1] Many authors use the word *glide-reflection* for transflection.

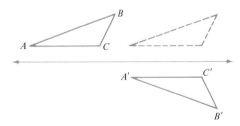

It is possible to show that the set consisting of all translations, rotations, reflections, and transflections forms a group under the operation of composition of transformations. Although the argument to establish this is not difficult, it is rather tedious and laborious and not especially instructive, so we omit it.

During the sequence of considering translations, rotations, reflections, and transflections of a plane α, we showed that each of these is a rigid motion of α. The converse is also true. That is, *translations, rotations, reflections*, and *transflections are the only rigid motions!* One often sees reference to "the group of rigid motions." The group of rigid motions is the group which consists of the set of all translations, rotations, reflections, and transflections on a plane α and the operation of composition of transformations.

The concept of rigid motion can be used to define congruence.

Definition 10.8 Two geometric figures in a plane α are *congruent* if and only if there exists a rigid motion of α such that the one figure is the image set of the other under the rigid motion.

In Section 10.7, we point out the great importance of Definition 10.8 for the prospective elementary teacher.

Exercise set 10.4

1. Let t denote a translation of a plane α along a line $\mathscr{L}$ and f denote a reflection of α in $\mathscr{L}$.

   a. By construction, show that the inverse of $f \circ t$ is $t^{-1} \circ f^{-1}$, where t^{-1} and f^{-1} denote the inverse of t and f, respectively.

   b. By construction, show that the inverse of $t \circ f$ is $f^{-1} \circ t^{-1}$.

2. By construction, determine the image of $\triangle ABC$ under a reflection in line $\mathscr{L}$ as in the accompanying model.

10 transformations of a plane—motion geometry

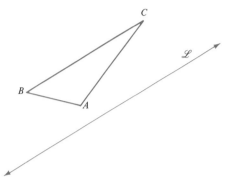

3. If r is a rotation of a plane α around a point P and f is a reflection in a line $\mathcal{L}$, by construction show that $f \circ r$ is a transflection of α. (*Hint:* Consider the image set of $\triangle ABC$ in α.)

4. Prove: If f is a reflection of a plane α, then the image of angle ABC in α is an angle which is congruent to $\angle ABC$.

5. Prove: If f is a reflection of a plane α, then the image of triangle ABC in α is a triangle which is congruent to $\triangle ABC$.

6. Draw models which suggest the image of $\triangle ABC$ suggested by the accompanying model under a transformation of the plane containing $\triangle ABC$ as follows:

a. A reflection in a line $\mathcal{L}$, $\mathcal{L} \cap \triangle ABC = \varnothing$ such that:
   (1) $\mathcal{L} \parallel \overleftrightarrow{CB}$,   (2) $\mathcal{L} \parallel \overleftrightarrow{BA}$,   (3) $\mathcal{L} \parallel \overleftrightarrow{AC}$.

b. A reflection in a line $\mathcal{L}$ such that $C \in \mathcal{L}$ and $\mathcal{L} \perp \overleftrightarrow{AC}$.

c. A reflection in $\overleftrightarrow{AB}$ followed by:
   (1) a reflection in $\overleftrightarrow{A'C'}$,
   (2) a reflection in $\overleftrightarrow{B'C'}$,
   (3) a clockwise rotation around C' with measure $^+30$,
   (4) a clockwise rotation around A' with measure $^+60$,
   (5) a translation along $\overleftrightarrow{B'C'}$,
   (6) a translation along $\overleftrightarrow{AB}$.

7. The following models are of triangles such that $\triangle ABC$ is scalene, $\triangle DEF$ is isosceles, $\triangle GHI$ is equilateral.

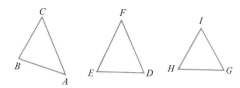

a. Are there any rigid motions of the plane, other than the identity, which map $\triangle ABC$ onto itself? A cutout and a drawing may help. If you discover any, describe them carefully.

b. Are there any rigid motions of the plane, other than the identity, which map $\triangle DEF$ onto itself? If you discover any, describe them carefully.

c. Are there any rigid motions of the plane, other than the identity, which map $\triangle GHI$ onto itself? If you discover any, describe them carefully.

10.6 SYMMETRY

The beauty of many natural objects, artifacts, and contemporary art works results from their symmetry. The word "symmetry" has something in common with the word "curve"; we use both of them intuitively without knowing their precise definition; both are abstractly definable in terms of the concept of function. There are many kinds of symmetries; we will limit our discussion to two kinds, reflection symmetries and rotation symmetries.

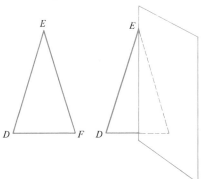

FIG. 10.15
Model to illustrate a reflection symmetry.

The intuitive idea of a reflection symmetry is easily illustrated by means of a model of an isosceles triangle with a mirror placed so it forms an altitude to the side not congruent to the other two as illustrated in Fig. 10.15. One observes that the image of "one-half" of the isosceles triangle looks exactly like "the other half." The geometric abstraction of a mirror is a line of reflection; therefore, an isosceles triangle has symmetry because there is a reflection of the plane, other than the identity, under which the triangle is a fixed set.

Next consider the model of a curve shown in Fig. 10.16. Some thought (indeed, try a mirror!) quickly convinces us that the model does have a reflection symmetry; moreover, if the model is "given a half-turn" around the point P, it is apparent that the model will "fall back on itself." We conclude, therefore, that the model in Fig. 10.16 has a rotation symmetry.

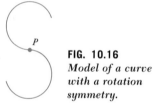

FIG. 10.16
*Model of a curve
with a rotation
symmetry.*

The general idea of symmetry is given in Definition 10.9.

Definition 10.9 A plane figure $\mathscr{F}$ has *symmetry* if and only if there exists a rigid motion t, not the identity, of the plane such that $\mathscr{F}$ is a fixed set under t. In this case, $\mathscr{F}$ is said to have t-symmetry. If t is a reflection, then $\mathscr{F}$ is said to have a *reflection symmetry*. If t is a rotation, then $\mathscr{F}$ is said to have a *rotation symmetry*.

If a plane figure $\mathscr{F}$ has a reflection symmetry, we often say it has a line of symmetry, or has a "folding" or a "flipping" symmetry. If a plane figure $\mathscr{F}$ has a rotation symmetry, we often say it has a point of symmetry or a "turning" symmetry.

Exercise set 10.5

1. Determine each turning and folding symmetry of each of the accompanying models by "drawing" the line or point of symmetry. Indicate the number of each type of symmetry for each figure.

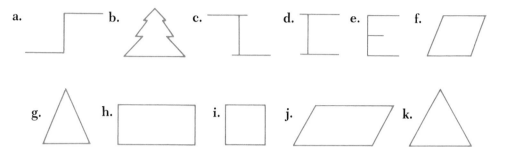

a.   b.   c.   d.   e.   f.

g.   h.   i.   j.   k.

10.6.1   The group of symmetries of an equilateral triangle—another look at transformation groups

Consider the accompanying model of an equilateral triangle ABC. The point P which is the intersection of the three altitudes of $\triangle ABC$ is a point of symmetry of $\triangle ABC$. A rotation of the plane

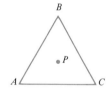

around P with measure $^+120$, denoted by r_{120}, maps $\triangle ABC$ onto itself as suggested by the following model. That is, under the rotation of the plane around P, we have A mapped to B, B to C, and C to A. A triangle is completely determined by its vertices, therefore, the image of $\triangle ABC$ under a rigid motion of the plane is completely described by indicating the image points of its vertices. If we use the notation introduced in Chapter 9, p. 228, the result of applying the rotation of the plane around point P to $\triangle ABC$ can be denoted by $\begin{pmatrix} ABC \\ BCA \end{pmatrix}$. That is, the clockwise ro-

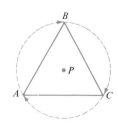

tation r_{120} determines one of the transformations of $\{A,B,C\}$. Do you recall that there are exactly six transformations of $\{A,B,C\}$? Let us investigate to determine whether there is a rigid motion of the plane which determines (or corresponds to) each of these transformations. We leave to the reader the problem of showing that r_{240}, the clockwise rotation of the plane around P with measure $^+240$, determines the transformation denoted by $\begin{pmatrix} ABC \\ CAB \end{pmatrix}$ and that $r_{360} = r_0$ is the identity transformation $\begin{pmatrix} ABC \\ ABC \end{pmatrix}$.

Now the two rotation symmetries of $\triangle ABC$ together with the identity transformation determine three transformations of $\{A,B,C\}$. It is intuitively clear that $\triangle ABC$ has no more rotation symmetries. Next we consider reflection (or folding or flipping) symmetries. Refer to the accompanying model to visualize the image of $\triangle ABC$ under the reflection of the plane in the line which contains the altitude of $\triangle ABC$ through A. It is easy to see that this reflection determines the transformation of $\{A,B,C\}$ denoted by $\begin{pmatrix} ABC \\ ACB \end{pmatrix}$. The reflections of the plane in the lines containing the other two alti-

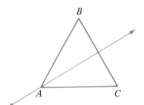

tudes of $\triangle ABC$ determine the transformations of $\{A,B,C\}$ denoted by $\begin{pmatrix} ABC \\ CBA \end{pmatrix}$ and $\begin{pmatrix} ABC \\ BAC \end{pmatrix}$. (Which one of the two is the reflection in the line which contains the altitude through B? through C?) It is intuitively clear that $\triangle ABC$ has no more reflection symmetries and indeed that $\triangle ABC$ has only the five symmetries described, which together with the identity transformation constitute the six transformations of $\{A,B,C\}$.

A model of an equilateral triangle $\triangle ABC$ can be used to diagram the transformations of $\{A,B,C\}$ and to determine the composition of transformations. We make the following agreements:

1. The accompanying model is the "home position" of $\triangle ABC$.

2. r_0, r_{120}, and r_{240} denote clockwise rotations with measures 0, $^+120$, and $^+240$, respectively.
3. f_1, f_2, and f_3 are reflections of the plane in lines $\mathcal{L}_1$, $\mathcal{L}_2$, and $\mathcal{L}_3$, respectively, as suggested in the accompanying model.

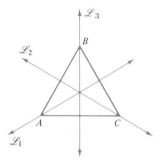

Using these agreements, we illustrate some of the transformations of $\{A,B,C\}$ (or symmetries of $\triangle ABC$). If we concentrate on the A, B, and C as "locations," then the accompanying diagram

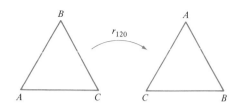

suggests that as a result of the rotation r_{120}, "A is where B was," "B is where C was," and "C is where A was"; that is, $A \rightarrow B$, $B \rightarrow C$, and $C \rightarrow A$; therefore, this diagram illustrates the transformation of $\{A,B,C\}$ (or symmetry of $\triangle ABC$) denoted by $\begin{pmatrix} ABC \\ BCA \end{pmatrix}$.

The accompanying diagram illustrates the transformation (or

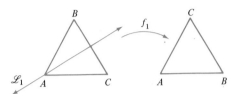

symmetry of $\triangle ABC$) denoted by $\begin{pmatrix} ABC \\ A\,?\,? \end{pmatrix}$. We leave to the reader the problem of completing the following diagram. This is the

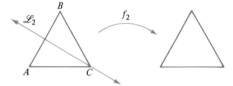

diagram which illustrates the transformation of $\{A,B,C\}$ (or symmetry of $\triangle ABC$) denoted by $\begin{pmatrix} ABC \\ ?\,?\,? \end{pmatrix}$.

The next diagrams illustrate how a sequence of such diagrams can be used to determine composition of transformations of $\{A,B,C\}$.

We determine $r_{120} \circ f_2$ by using the accompanying model. Therefore, $r_{120} \circ f_2 = f_3$ or $\begin{pmatrix} ABC \\ BCA \end{pmatrix} \circ \begin{pmatrix} ABC \\ BAC \end{pmatrix} = \begin{pmatrix} ABC \\ CBA \end{pmatrix}$.

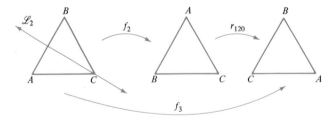

We leave it for the reader to determine $f_2 \circ r_{120}$, which he can do by completing the accompanying diagram. Therefore, $f_2 \circ r_{120} = ?$.

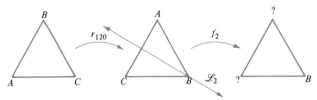

The set $S_3 = \{r_0, r_{120}, r_{240}, f_1, f_2, f_3\}$ includes the set of symmetries of an equilateral triangle. Exercise 1 of Exercise Set 10.6 gives the reader the opportunity to show that S_3, $\circ$ is a group; it is called the "group of symmetries of an equilateral triangle," or "the symmetric group on three elements."

Exercise set 10.6

1. Let $S_3 = \{r_0, r_{120}, r_{240}, f_1, f_2, f_3\}$ be the five symmetries of an equilateral triangle ABC together with the identity transformation. Let $\circ$ be the operation of composition of transformations.

   a. Complete a table for $\circ$ on S_3.
   b. Is S_3 closed with respect to $\circ$?
   c. Is $\circ$ well defined on S_3?
   d. Is $\circ$ a binary operation on S? Explain.
   e. Prove the following statement.

   There exists $x \in S_3$ such that for every $y \in S_3$ $x \circ y = y$ and $y \circ x = y$.

   f. Prove the following statement.

   For every $y \in S_3$ there exists $z \in S_3$ such that $z \circ y = x$ and $y \circ z = x$,

   where x is the element you determined in part e.
   g. Is $\circ$ associative on S_3?
   h. Is S_3, $\circ$ a group?
   i. Is S_3, $\circ$ a commutative group? Explain.

2. The group of symmetries of a square, a "subgroup" of the group of transformations on a set of four elements (or the symmetric group on four elements), can be developed by considering the rotation and reflection (turning and folding) symmetries of a square.

   Let r_0, r_{90}, r_{180}, r_{270}, d_1, d_2, v, and h denote the rigid motions of a plane which determine motions of square $ABCD$, or transformations of $\{A,B,C,D\}$ as in the accompanying models.

$$A \to A$$
$$B \to B$$
$$C \to C$$
$$D \to D$$

$$\begin{pmatrix} ABCD \\ ABCD \end{pmatrix}$$

$$A \to B$$
$$B \to C$$
$$C \to D$$
$$D \to A$$

$$\begin{pmatrix} ABCD \\ BCDA \end{pmatrix}$$

$$A \to \text{?}$$
$$B \to \text{?}$$
$$C \to \text{?}$$
$$D \to \text{?}$$

$$\begin{pmatrix} ABCD \\ \text{?}\,\text{?}\,\text{?}\,\text{?} \end{pmatrix}$$

$$A \to \text{?}$$
$$B \to \text{?}$$
$$C \to \text{?}$$
$$D \to \text{?}$$

$$\begin{pmatrix} ABCD \\ \text{?}\,\text{?}\,\text{?}\,\text{?} \end{pmatrix}$$

$$A \to \text{?}$$
$$B \to \text{?}$$
$$C \to \text{?}$$
$$D \to \text{?}$$

$$\begin{pmatrix} ABCD \\ \text{?}\,\text{?}\,\text{?}\,\text{?} \end{pmatrix}$$

$$A \to \text{?}$$
$$B \to \text{?}$$
$$C \to \text{?}$$
$$D \to \text{?}$$

$$\begin{pmatrix} ABCD \\ \text{?}\,\text{?}\,\text{?}\,\text{?} \end{pmatrix}$$

$$A \to \text{?}$$
$$B \to \text{?}$$
$$C \to \text{?}$$
$$D \to \text{?}$$

$$\begin{pmatrix} ABCD \\ \text{?}\,\text{?}\,\text{?}\,\text{?} \end{pmatrix}$$

$$A \to \text{?}$$
$$B \to \text{?}$$
$$C \to \text{?}$$
$$D \to \text{?}$$

$$\begin{pmatrix} ABCD \\ \text{?}\,\text{?}\,\text{?}\,\text{?} \end{pmatrix}$$

Let $S_4 = \{r_0, r_{90}, r_{180}, r_{270}, d_1, d_2, v, h\}$ and let $\circ$ be defined on S_4 as usual for compositions of transformations.   We illustrate how "motions" of the square can be used to determine composition of transformations of $\{A,B,C,D\}$.   We determine $d_2 \circ r_{270}$ or,
$$\begin{pmatrix} ABCD \\ ADCB \end{pmatrix} \circ \begin{pmatrix} ABCD \\ BCDA \end{pmatrix} = \begin{pmatrix} ABCD \\ DCBA \end{pmatrix}$$ as in the accompanying model.

Therefore, $d_2 \circ r_{270} = v$.

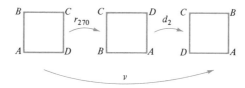

a. Complete the table for $\circ$.
b. Is S_4 closed with respect to $\circ$?
c. Is $\circ$ well defined on S_4?
d. What is the identity in S_4 for $\circ$?
e. Prove that every element of S_4 has an inverse with respect to $\circ$.
f. Is $\circ$ associative on S_4? Explain.
g. Is $S_4, \circ$ a group? Explain.
h. Is $S_4, \circ$ a commutative group? Explain.

10.6.2 Classification of plane figures according to their symmetries

The two models of geometric figures in Fig. 10.17 clearly illustrate that the number of symmetries which a plane figure has does

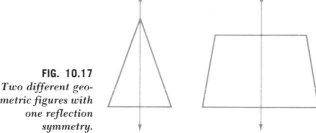

FIG. 10.17
Two different geometric figures with one reflection symmetry.

not uniquely determine the type of figure it is. Apparently, both an isosceles triangle and an "isosceles" trapezoid have exactly one symmetry—a reflection symmetry. On the other hand, the number of symmetries of polygons already classified according to the number of sides does give further classification of the polygons.

A triangle is scalene if and only if it has zero lines of symmetry.
A triangle is isosceles if and only if it has one line of symmetry.
A triangle is equilateral if and only if it has three lines of symmetry.

It is interesting that a classification of triangles according to the number of line symmetries agrees with the more familiar way of classifying them according to the congruence of their sides.   That is, zero lines of symmetry—zero sides congruent, one line of symmetry—one pair of sides congruent, three lines of symmetry— three pairs of sides congruent.   Such is not the case with quadrilaterals.   A quadrilateral can have zero, one, two, or four lines of symmetry as the models in Fig. 10.18 illustrate.

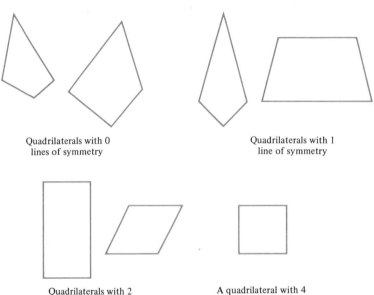

Quadrilaterals with 0
lines of symmetry

Quadrilaterals with 1
line of symmetry

Quadrilaterals with 2
lines of symmetry

A quadrilateral with 4
lines of symmetry

FIG. 10.18
*Quadrilaterals
with* (a) *zero,* (b)
one, (c) *two, and*
(d) *four lines of
symmetry.*

In doing Exercise 7 of Exercise Set 10.4, p. 263, the reader drew models to suggest that the only triangle with a rotation (or turning) symmetry is equilateral and that it has two rotation symmetries.   A rectangle has one rotation symmetry and a square has three.   The set of rotation symmetries of a circle is infinite; also, the set of reflection symmetries of a circle is infinite.

10.7 OVERVIEW

The object of this chapter was to define the geometric analogs of picking up, sliding, turning, flipping, or whatever movements one can do to physical objects without changing their size or shape.

We "mathematized" the problem by defining the idea of a rigid motion. This mathematizing of the problem led to the discovery that there are really only four different ways to move an object without changing its size or shape! The four movements are slides, turns, flips, and a slide followed by a flip.

A rigid motion of a plane α is a transformation of α which preserves the distance between points. The four distinct rigid motions can be characterized by their fixed points and fixed sets as follows.

A *translation* has *zero fixed points*, and all the lines parallel to the line of translation are fixed sets.

A *rotation* has *one fixed point*, and all concentric circles with the same center as the rotation are fixed sets.

A *reflection* has *infinitely many fixed points*, and all lines perpendicular to the line of reflection are fixed sets.

A *transflection* has *zero fixed points*, and *one line as a fixed set*.

For the elementary school teacher the really significant outcome of the study of the rigid motions of a plane lies in the fact that two coplanar figures are congruent if one is the image of the other under some rigid motion of the plane. What this does is to provide a mathematical framework with tremendous flexibility for treating the ideas of congruence and symmetry intuitively through the use of cut-out models of geometric figures. The teacher can be confident in talking about moving, folding, turning, and flipping models that he is developing geometric ideas which are consistent with those his students may later study more formally.

Some exercises in the Miscellaneous Exercises suggest some activities which a teacher can use to aid his students in the discovery of many geometric concepts; these activities are really justifiable because of *transformation geometry*.

Miscellaneous exercises

1. Without using Theorem 10.1 prove the following:

   a. Translations of a plane preserve betweenness of points in the plane.

   b. Rotations of a plane preserve betweenness of points in the plane.

    c. Reflections of a plane preserve betweenness of points in the plane.

2. Consider the accompanying model union-of-line segments curve.   Draw a model which suggests its image under:

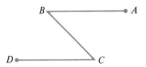

    a. a reflection of the plane in $\overleftrightarrow{BC}$

    b. a clockwise rotation around D with measure $^{+}90$

    c. a counterclockwise rotation around the midpoint of $\overline{BC}$ with measure $^{+}180$

    d. a clockwise rotation around the midpoint of $\overline{BC}$ with measure $^{+}180$

    e. a translation along $\overleftrightarrow{CD}$ with measure CD followed by a reflection in $\overleftrightarrow{CD}$

    f. a reflection in the perpendicular bisector of $\overline{AB}$ followed by a reflection in the line parallel to $\overline{AB}$ and $\overline{CD}$ which passes through the midpoint of $\overline{BC}$

    g. a reflection in $\overleftrightarrow{DQ}$, where $\overleftrightarrow{DQ} \parallel \overleftrightarrow{CB}$ followed by a clockwise rotation around A' with measure $^{+}45$

    h. a clockwise rotation around D with measure $^{+}90$ followed by a clockwise rotation around B' with measure $^{+}90$.

3. Prove the following:

    a. If t is a translation of a plane α, then the image of $\angle ABC$ under t *is an angle* and is *congruent* to $\angle ABC$.

    b. If r is a rotation of a plane α, then the image under r *is an angle* and is *congruent* to $\angle ABC$.

    c. If f is a reflection of a plane α, then the image under f *is an angle* and is *congruent* to $\angle ABC$.

Note: The following exercises suggest activities which one can carry out with paper models to lead elementary school students to discover geometric relationships.

4. Using any sheet of paper (as a model of a plane):

    a. Fold it to form a right angle.
    b. Fold it to form a rectangle.
    c. Determine, by folding, the *point* of symmetry of the rectangle.
    d. Fold the rectangle to form a square.
    e. Determine, by folding, the point of symmetry of the square.

5. "Draw a line segment AB" on a sheet of paper and fold it to determine its midpoint.

6. Cut and fold paper to construct:

    a. an isosceles triangle,
    b. an equilateral triangle.

7. Construct and cut out a paper model of an equilateral triangle ABC.

    a. Determine, by folding, the line of symmetry through A and the point P of symmetry of $\triangle ABC$.
    b. Determine, by folding, a line through P parallel to the base $\overline{BC}$.
    c. Determine, by folding, the ratio $\dfrac{AP}{PQ}$, where Q is the point where the altitude from A intersects $\overline{BC}$.

8. Cut a model of a scalene triangular region as illustrated in the accompanying model.

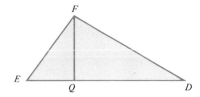

    a. Determine, by folding, the altitude from F, call it $\overline{FQ}$.
    b. Make folds so that E, F, and D "all fall on Q."
    c. What conclusion can you draw?

9. Cut and fold a sheet of paper to construct an isosceles trapezoid.

10. Cut out a paper model of a region whose boundary is a parallelogram. Let b be the measure of the base and h the measure of the altitude of the parallelogram. Then, cut the region into two parts and move the two parts to show that the measure of this region is equal to the measure of a rectangular region whose base has measure b and altitude has measure h.

11. Cut out a paper model of a trapezoidal region. Then, cut and move the parts to show that the measure of a trapezoidal region is $\frac{+1}{+2}(b_1 + b_2)h$.

Cut→   Rotate

12. Consider quadrilateral $ABCD$.
    a. Fold to determine two triangles.
    b. Fold further to determine the sum of the measures of the angles of quadrilateral $ABCD$.
    c. Do the same for a pentagon, hexagon, heptagon, octagon.
    d. Can you make a generalization about the sum of the measures of the angles of a polygon?

13. How many lines and points of symmetry are there of the following:
    a. a regular pentagon?
    b. a regular hexagon?
    c. a regular heptagon?
    d. a regular octagon?

14. Determine, by folding, several lines of symmetry of a circle. What do you observe?

15. Cut out a paper model of a scalene triangle. By folding, determine the center of the circle which contains the vertices of the triangle.

answers and suggestions for selected exercises

1. **a.** The set of all whole numbers between 0 and 6
   c. The set of all whole numbers between 0 and 106
   e. The set of all whole numbers between 3 and 10

2. **a.** $\varnothing$ **c.** $\{6,7\}$

3. **a.** $\{x | x$ is a whole number greater than 0 and x is a multiple of 10$\}$
   c. $\{x | x$ is a whole number and $4 < x < 11\}$

1. Jack $\leftrightarrow$ Bob
   Mary $\leftrightarrow$ Jack
   Sue $\leftrightarrow$ June

3. **a.** True   **b.** False

4. **a.** True   **c.** False   **e.** True
   g. False   **i.** False   **k.** True
   m. False   **o.** True   **q.** False

5. **a.** True   **c.** True   **e.** True   **g.** True
   i. False   **k.** False   **m.** False

6. **a.** True **c.** True **e.** True

7. ∅;
{a}, ∅;
{a}, {b}, {a,b}, ∅;
{a}, {b}, {c}, {a,b}, {a,c}, {b,c}, {a,b,c}, ∅;
{a}, {b}, {c}, {d}, {a,b}, {a,c}, {a,d}, {b,c}, {b,d}, {c,d},
{a,b,c}, {a,b,d}, {a,c,d}, {b,c,d}, {a,b,c,d}, ∅;
{a}, {b}, {c}, {d}, {e}, {a,b}, {a,c}, {a,d}, {a,e}, {b,c}, {b,d},
{b,e}, {c,d}, {c,e}, {d,e}, {a,b,c}, {a,b,d}, {a,b,e}, {a,c,d},
{a,c,e}, {a,d,e}, {b,c,d}, {b,c,e}, {b,d,e}, {c,d,e}, {a,b,c,d},
{a,b,c,e}, {a,b,d,e}, {a,c,d,e}, {b,c,d,e}, {a,b,c,d,e}, ∅

8. 256

10. **a.** True **b.** False **c.** True

12. **a.**

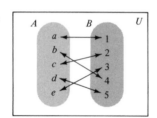

Exercise set 1.4, p. 15

1. **a.** Not a statement
c. Not a statement
e. Not a statement
g. Not a statement

2. **a.** Bill is old and people are funny. *(F)*
Bill is old or people are funny. *(T)*
c. 4 + 3 = 9 and 7 + 3 = 12. *(F)*
4 + 3 = 9 or 7 + 3 = 12. *(F)*

3. **b.** Bill is not old or people are not funny. *(T)*
Bill is not old and people are not funny. *(F)*
c. 4 + 3 ≠ 9 or 7 + 3 ≠ 12. *(T)*
4 + 3 ≠ 9 and 7 + 3 ≠ 12. *(T)*

4.

p	q	*not p*	*not q*	*p or q*	*(not p) and (not q)*
T	T	F	F	T	F
T	F	F	T	T	F
F	T	T	F	T	F
F	F	T	T	F	T

Exercise set 1.5, p. 19

1. (1) **a.** If London is a city, then Venus is a planet. (*T*)
       b. If Venus is a planet, then London is a city. (*T*)
       c. London is a city if and only if Venus is a planet. (*T*)
    (3) **a.** If elephants are small, then monkeys are the predecessors of man. (*T*)
       b. If monkeys are the predecessors of man, then elephants are small. (*F*)
       c. Elephants are small if and only if monkeys are the predecessors of man. (*F*)

2. **a.** If a number is divisible by 4, then it is divisible by 2.
    c. If the negation of the statement is true, then the statement is false.

Exercise set 1.6, p. 23

1. **a.** (1) {5,10,15,20}   (3) {10,15}
       (5) Y   (7) X
       (9) {1,3,5,7, . . .}
       (11) {(10,10),(10,15),(15,10),(15,15)}
       (13) {20}   (15) {1,2,3,4,6,7,8,9,11,12,13,14}
    b. (1) True   (3) True   (5) False   (7) True
       (9) True   (11) False   (13) False   (15) False

2. **a.** {a,b,c}   **c.** $\varnothing$ **e.** $\varnothing$

3. **a.** {1,7,8,9,10}   **c.** U   **e.** U   **g.** $\varnothing$
    i. {1,2,3,7,8,9,10}   **k.** {1,8,9,10}   **m.** U   **o.** U

4. **a.** B,C   **c.** $6 \in A$ and $6 \in B$ **e.** $9 \in M$ and $9 \in N$
    g. $R \cap S$ **i.** $B \cup C$

Exercise set 1.7, p. 29

1. **a.** $\{(0,1),(0,2),(0,3),(0,4),(1,2),(1,3),(1,4),(2,3),(2,4),(3,4)\}$
    c. $\{(1,4),(2,3),(3,2),(4,1)\}$

3. **a.** $\{(a,c),(a,d),(a,e),(b,c),(b,d),(b,e)\}$
    c. $\{(c,c),(c,d),(c,e),(d,c),(d,d),(d,e),(e,c),(e,d),(e,e)\}$
    e. $R_1 = \{(a,c),(a,d),(a,e)\}$.
    $R_2 = \{(a,c)\}$.
    $R_3 = \varnothing$.
    g. $R_1 = \{(a,a)\}$.
    $R_2 = \{(a,a),(b,b)\}$.
    $R_3 = A \times A$.

4. R_1: $\{1\}$; $\{9,10,11\}$
    R_3: $\{1,2,3\}$; $\{9,10,11\}$

5. **a.** For every x, $y \in C$, $x \, R \, y$ if and only if $y = x + 1$.
    $R = \{(x,y) \in C \times C \,|\, y = x + 1\}$.
    c. For every x, $y \in C$, $x \, Q \, y$ if and only if x is an even counting number and $y = x + 3$.
    $Q = \{(x,y) \in C \times C \,|\, x$ is an even counting number and $y = x + 3\}$.
    e. For every x, $y \in C$, $x \, N \, y$ if and only if $x + y = 9$.
    $N = \{(x,y) \in C \times C \,|\, x + y = 9\}$.

6. **a.** $R_1 = \{(1,1),(2,2),(3,3),(4,4),(5,5)\}$.
    c. $R_3 = \{(1,2),(2,4),(3,6),(4,8)\}$.

7. **a.** $\{(a,c),(a,d),(b,c),(b,d)\}$
    b. $\{((a,c),1),((a,d),1),((b,c),1),((b,d),1)\}$
    c. $\{((a,c),1),((a,d),1)\}$

8. **a.** One-to-many   **c.** Many-to-one

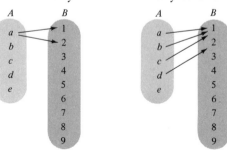

e. One-to-many   **g.** One-to-many

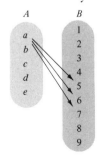 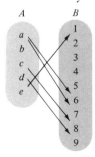

Exercise set 1.8, p. 32

1. a. The relation "is the brother of" defined on S, where S is the set of people in the United States, can be considered to have the reflexive property with an "appropriate" definition of "is the brother of."   Transitive

    c. Reflexive, symmetric, transitive

2. Subset: reflexive, transitive
    One-to-one correspondence: reflexive, symmetric, transitive

3. a. $:= \{(1,1),(1,41),(1,111),(1,301),(41,1),(41,41),(41,111),$
    $(41,301),(111,1),(111,41),(111,111),(111,301),(301,1),$
    $(301,41),(301,111),(301,301),(2,2),(2,22),(2,42),$
    $(2,15732),(22,2),(22,22),(22,42),(22,15732),(42,2),$
    $(42,22),(42,42),(42,15732),(15732,2),(15732,22),$
    $(15732,42),(15732,15732),(3,403),(403,3),(4,4),(4,44),$
    $(4,444),(4,974),(4,16234),(44,4),(44,44),(44,444),$
    $(444,444),(444,974),(444,16234),(974,4),(974,44),$
    $(974,444),(974,974),(974,16234),(16234,4),(16234,44),$
    $(16234,444),(16234,974),(16234,16234),(5,5)\}.$

    b. The relation $:$ on Q is reflexive, symmetric, and transitive; hence, it is an equivalence relation.

    c. (1) True   (2) False
        (3) True   (4) False

5. a. Reflexive, symmetric, transitive
    b. (1) True   (2) False
        (3) True   (4) False
        (5) False

Exercise set 1.9, p. 37

1. a. One-to-one correspondence, relation from A to B, function from A to B
   c. Many-to-one, relation from A to B
   e. One-to-many, relation from A to B
   g. One-to-one correspondence
   i. Relation from A to B

2. a.   **c.**

   e.   **g.**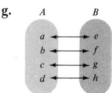

   i.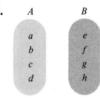

3. a. $\{(a,a),(a,b),(b,a),(b,b)\}$ **b.** Yes
   c. $\{((a,a),e),((a,b),e),((b,a),e),((b,b),e)\}$

5. a. Yes **c.** No **e.** No

6. a. Yes **b.** 4, 16, 36, 64 **c.** x^2
   d. $f = \{(x,f(x)) \in X \times Y | f(x) = x^2\}$ or
   $f = \{(x,y) \in X \times Y | y = x^2\}$.
   e. $\{2,4,6,8\}$ **f.** $\{4,16,36,64\}$

7. a. $g(3) = (3)^2 + 2 = 11$. **c.** 18 **e.** 38 **g.** 66
   h. (1) False (3) False
   i. (1) 11 (3) 9 (5) 13

ANSWERS AND SUGGESTIONS FOR SELECTED EXERCISES

Exercise set 2.1, p. 49

1.　**a.**　PF0.　Every row is a subset of S.
　　　　PF1.　Every two rows have exactly one tree in common.
　　　　PF2.　Every tree in S is in exactly two rows.
　　　　PF3.　There are exactly four rows in S.

　　b.

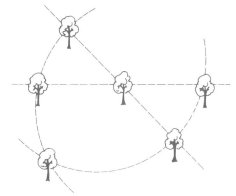

　　c.　Orchard
　　d.　**1.**　Not all rows contain the same tree.
　　　　2.　There are exactly six trees in the orchard.
　　　　3.　Every row contains exactly three trees.

3.　**a.**

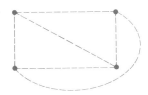

　　b.　Proof

1.	There are at least two points, call them A and B.	**1.** Postulate 1
2.	A and B are on exactly one line, call it l.	**2.** Definition 2, Step 1
3.	There exists a point, call it C, not on line l.	**3.** Postulate 2
4.	There exists a line, call it m, that is parallel to l and contains C.	**4.** Postulate 3

5. Line m contains two points, call them C and D.

    5. Definition 2

6. Therefore, there exists at least four points.

    6. Steps 1, 4, and 5

7. Assume that there are more than four points; that is, assume that there is at least one point E which is distinct from A, B, C, and D.

    7. Assumption

8. E is not on lines l or m.

    8. Definition 2, Steps 1, 5

9. There exists a line, call it n, which contains points E and D.

    9. Definition 2

10. There exists a line, call it p, which contains points E and C.

    10. Definition 2

11. Line n is parallel to line l; line p is parallel to line l.

    11. Definition 3

12. Therefore, there exists two lines p and n which contain point E that are parallel to line l.

    12. Steps 9, 10, 11

13. The assumption that there are more than four points leads to the contradiction that there exists more than one line parallel to a given line containing a given point; therefore, the assumption that there are more than four points is false.

14. There are exactly four points in S.

    14. Why?

c. Direct proof

1. There are exactly four points in S, call them A, B, C, and D.

    1. Exercise 3b

2. A and B are on a line—l. C and D are on a line—m. B and D are on a line—n.

    2. Definition 2

A and D are on a line—o.
C and B are on a line—p.
A and C are on a line—q.

3. Therefore,
A is on l, o, and q;
B is on l, n, and p;
C is on m, p, and q;
D is on m, n, and o.

3. Step 2

Proof by contradiction

1. Assume that there is a point, say A, which is on less than two lines; that is, A is on zero lines or on one line.

1. Assumption

2. There are at least two points, say A and B.

2. Postulate 1

3. $\{A,B\}$ is a line, say $\mathscr{L}$.

3. Definition 2

4. There is a point, say C, not on $\mathscr{L}$.

4. Postulate 2

5. $\{A,C\}$ is a line, say $\mathscr{M}$.

5. Definition 2

6. $\therefore$ A is on at least two lines.

6. Steps 3 and 5

7. The assumption that there is a point A which is on less than two lines leads to the contradiction that A is on less than two lines and is on at least two lines; therefore, the assumption that there is a point which is on less than two lines is false, and hence, every point is contained on at least two lines.

4. a. Proof

1. Every two wires have at least one bead in common.

1. Postulate 1

2. Every two wires have at most one bead in common.

2. Postulate 2

3. Therefore, every two wires have exactly one bead in common.

3. Trichotomy postulate

   e. Three

5. **a.** Six

   b. Let the politicians be A, B, C, D, E, and F.

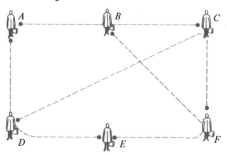

   c. No

   d. No. If a committee has less than three politicians, then there would exist two committees without a politician in common. This would contradict Postulate 3.

Exercise set 2.2, p. 60

1. Proof

1. S is an infinite set of points. Call two of the points A and B.	**1.** Postulate 2.1
2. There exists exactly one line which contains A and B, call it $\mathscr{L}$.	**2.** Theorem 2.1
3. There exists a point $P \in S$ such that P is not on $\mathscr{L}$.	**3.** Postulate 2.5
4. Therefore, $\{A,B,P\}$ is a noncollinear set.	**4.** Definition of noncollinear set

2. **a.**

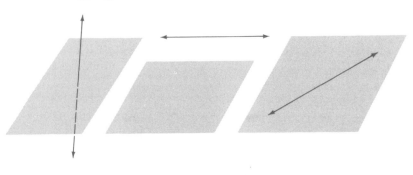

b. Outline of proof

1. $\mathcal{L} \cap \alpha$ contains exactly one point or $\mathcal{L} \cap \alpha$ does not contain exactly one point.
2. Assume $\mathcal{L} \cap \alpha$ does not contain exactly one point.
3. $\therefore \mathcal{L} \cap \alpha = \varnothing$ or $\mathcal{L} \cap \alpha$ contains at least two points.
4. If $\mathcal{L} \cap \alpha = \varnothing$, then the theorem is proved; therefore, assume $\mathcal{L} \cap \alpha$ contains at least two points.
5. $\therefore \mathcal{L} \subseteq \alpha$.
6. $\therefore \mathcal{L} \cap \alpha = \mathcal{L}$.

3. Proof

1. A, B, and C are noncollinear points.
2. There exists a plane, call it α, such that $\{A,B,C\} \subseteq \alpha$.
3. Assume there is another plane β, $\beta \neq \alpha$, such that $\{A,B,C\} \subseteq \beta$.
4. $\therefore \{A,B,C\} \subseteq \beta \cap \alpha$.
5. $\beta \cap \alpha = \varnothing$ or $\beta \cap \alpha = \mathcal{L}$.
6. $\therefore A,B,C \in \mathcal{L}$, and hence are collinear points.
7. The assumption that there is more than one plane which contains A, B, and C leads to the contradiction that A, B, and C are collinear and A, B, and C are not collinear; therefore, we must conclude that there is at most one plane which contains A, B, and C.
8. $\therefore$ There is exactly one plane which contains A, B, and C.

4. Proof

1. $A,B,C \in S$.
2. A, B, and C are collinear points or A, B, and C are noncollinear points.
3. Assume A, B, and C are collinear points.
4. $\therefore A$, B, and C are on a line $\mathcal{L}$.
5. There exists a point $P \in S$ such that $P \notin \mathcal{L}$.
6. $\therefore$ There exists a plane β which contains P and $\mathcal{L}$.
7. $\therefore A,B,C \in \alpha$.
8. Assume A, B, and C are noncollinear points.
9. $\therefore$ There exists a plane α which contains the points A, B, and C.

5. Proof

1. $\mathcal{L}$ and $\mathcal{M}$ are lines, $\mathcal{L} \neq \mathcal{M}$, $\mathcal{L} \cap \mathcal{M} \neq \varnothing$.

    2. $\therefore$ There exists a point P such that $\mathscr{L} \cap \mathscr{M} = \{P\}$.

    3. $\mathscr{M}$ is an infinite set; therefore, there exists a point $A \in \mathscr{M}$, such that $A \neq P$.

    4. There exists a plane β which contains $\mathscr{L}$ and A.

    5. $\{P,A\} \subseteq \beta$.

    6. $\therefore \beta$ contains $\mathscr{M}$.

    7. $\therefore \beta$ contains $\mathscr{L}$ and $\mathscr{M}$.

6. Proof

    1. α is a plane.

    2. α contains at least one line $\mathscr{L}$ and at least one point P not on $\mathscr{L}$.

    3. Line $\mathscr{L}$ is an infinite set of points.

    4. $\therefore$ Plane α is an infinite set of points.

7. Proof

    1. α and β are planes and $\alpha \cap \beta \neq \emptyset$.

    2. $\alpha \cap \beta = \emptyset$ or $\alpha \cap \beta$ is a line $\mathscr{L}$.

    3. $\therefore \alpha \cap \beta$ is a line $\mathscr{L}$.

    4. Line $\mathscr{L}$ is an infinite set of points.

    5. $\therefore \alpha \cap \beta$ is an infinite set of points.

8. Proof

    1. α is a plane.

    2. $\therefore \alpha$ is an infinite set of points; call two of them A and B.

    3. $\therefore$ There exists a line $\mathscr{L}$ such that $\{A,B\} \subseteq \mathscr{L}$.

    4. There exists a point $P \in \alpha$ not on $\mathscr{L}$.

    5. $\mathscr{L}$ is an infinite set of points.

    6. For every point X on $\mathscr{L}$ there exists a line which contains X and P.

    7. $\therefore \alpha$ contains an infinite set of lines.

9. Proof

    1. There exists a plane; call it α.

    2. α is an infinite set of points.

    3. For every line $\mathscr{M}$ on α, there exists a point P in S and not on α.

    4. For every line $\mathscr{M}$ on α, $\mathscr{M}$ and P determine a plane.

    5. α contains an infinite set of lines.

    6. $\therefore S$ contains an infinite set of planes.

ANSWERS AND SUGGESTIONS FOR SELECTED EXERCISES

Exercise set 3.1, p. 64

1. **a.** False   **c.** True   **e.** False   **g.** False
    i. False   **k.** False   **m.** True   **o.** True
    q. True   **s.** False   **u.** False   **w.** True
    y. True   **A.** True   **C.** False

2. **a.** $\varnothing$ **c.** $\overline{BC}$ **e.** $\overline{DG}$ **g.** $\{E\}$ **i.** $\{F\}$ **k.** $\overline{DG}$

5. **a.**

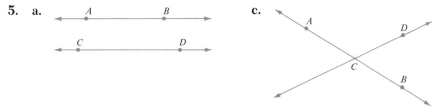

    e.   **g.**

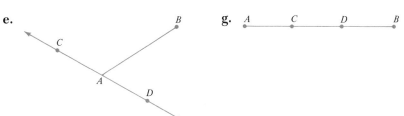

    i.   **k.**

    m.   **o.**

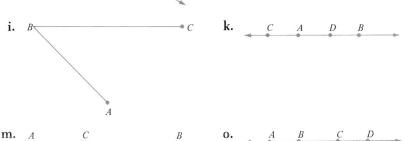

6. Proof

    1. A and B are two points in plane α.
    2. $\overleftrightarrow{AB} \subseteq \alpha$.   **3.** $\overline{AB} \subseteq \overleftrightarrow{AB}$.   **4.** $\therefore \overline{AB} \subseteq \alpha$.

Exercise set 3.2, p. 71

1. **a.** True   **c.** False   **e.** False   **g.** True
    i. True   **k.** True   **m.** True   **o.** True

2. **a.** False   **c.** False   **e.** False   **g.** True   **i.** True
   k. False   **m.** False   **o.** False   **q.** True

3. **a.** True   **c.** False   **e.** $\{C,F\}$
   g. $\varnothing$ **i.** False   **k.** $\varnothing$

4. Yes,

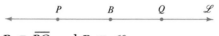

$B \in \overline{PQ}$ and $B \in \mathscr{L}$.
$\therefore B \in \overline{PQ} \cap \mathscr{L}$.

5. No

6. **a.**

B   C   D

c.

C A B D E F

e.

P   C
A   S
B
R

g.

B D C S

Exercise set 3.3, p. 77

1. **a.** True   **c.** False   **e.** True   **g.** True
   i. False   **k.** True   **m.** True   **o.** True
   q. False   **s.** True   **u.** True   **w.** False
   y. False   **A.** True   **C.** False   **E.** True

2. **a.** $\angle GDH$, $\angle GCF$, $\angle GCE$, $\angle HBE$, $\angle HBF$ **c.** $\{C,D\}$
   e. $\angle HBE$, $\angle GCE$ **g.** No   **i.** GCE, GCB
   k. $\overrightarrow{DG}$ and $\overrightarrow{DH}$ **m.** No   **o.** Yes, $\angle HBE$
   q. $\overline{DC} \setminus \{D\}$ **s.** $\overrightarrow{DH}$

4. **a.** One   **b.** No

5. **a.**

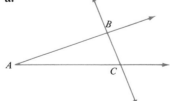

c.

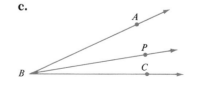

e.

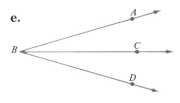

g.

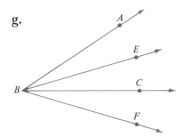

i.

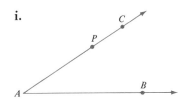

Exercise set 4.1, p. 84

1. **a.** Commutative property of multiplication
    c. Left distributive property of multiplication over addition
    e. Commutative property of addition
    g. Multiplication inequality transformation principle
    i. Multiplicative inverse property
    k. Multiplication inequality transformation principle
    m. Multiplicative inverse property
    o. Multiplication equation transformation principle

2. **a.** $^+5 < {}^+5$ or $^+5 = {}^+5$.

3. **a.** 1. $a,b \in \mathbb{R}$.
    2. $(a + b) + [(-a) + (-b)] = [a + (-a)] + [b + (-b)]$.
    (Several steps needed.)
    3. $= 0$.
    4. $\therefore (a + b) + [(-a) + (-b)] = 0$.
    5. $\therefore -(a + b) = (-a) + (-b)$.
    c. 1. $a,b \in \mathbb{R}$.
    2. Assume $a \geqslant b$.
    3. $\therefore a - b \geqslant 0$.
    4. $\therefore |a - b| = a - b$.
    5. Assume $a < b$.
    6. $\therefore a - b < 0$.
    7. $\therefore |a - b| = -(a - b)$.

    8. $-(a-b) = b-a$.

    9. $\therefore |a-b| = b-a$.

4. a. $^+5$ **c.** $^+11$ **e.** $^+6$

Exercise set 4.2, p. 90

1. a. $|^-1 - {}^-3| = {}^+2$ **c.** $^+2$ **e.** $^+4$ **g.** $^+3$

2. The proof that "For all $A,B,C \in S$, if A–B–C, then C–B–A" follows.

    1. $A,B,C \in S$.
    2. A–B–C.
    3. $AB + BC = AC$.
    4. $BC + AB = AC$.
    5. $BC = CB$, $AB = BA$, and $AC = CA$.
    6. $CB + BA = CA$.
    7. $\therefore C$–B–A.

The proof that "For all $A,B,C \in S$, if C–B–A, then A–B–C" is left to the reader.

4. a. (1) $|^+3\frac{1}{2} - {}^+1\frac{1}{2}| = {}^+2$.   (2) $^+1$ (3) $^+2$
     (4) Yes; $^+2 = {}^+2 \cdot {}^+1$.
  c. $^+2$

5. a. (1) $^+4\frac{1}{2}$ (2) $^+9$ (3) $\dfrac{^+1}{^+2}$ (4) $^+2$

     (5) Yes; $^+4\dfrac{1}{2} = \dfrac{^+1}{^+2} \times {}^+9$.

  b. (1) $^+3, {}^+6$, yes, $^+3 = \dfrac{^+1}{^+2} \times {}^+6$.

     (3) $^+1\dfrac{1}{2}, {}^+3$, yes, $^+1\dfrac{1}{2} = \dfrac{^+1}{^+2} \times {}^+3$.

     (5) $^+3, {}^+6$, yes, $^+3 = \dfrac{^+1}{^+2} \times {}^+6$.

Exercise set 4.3, p. 93

1. a. $|^+2 - {}^+4| = {}^+2$.   **c.** $|0 - {}^+2| = {}^+2$.   **e.** $|0 - {}^-1| = {}^+1$.

2. a. True, $m(\overline{AB}) = m(\overline{CD})$.   **c.** False, $\overline{BD} \cong \overline{DF}$.

  e. True, $\dfrac{^+1}{^+2} = \dfrac{^+1}{^+2}$.   **g.** True, $m(\overline{DE}) = m(\overline{ED})$.

i. True, $^+1 = \dfrac{^+1}{^+3} \times {}^+3$.

k. False, $\overline{BC}$ is not the same line segment as $\overline{DE}$.

m. True, $^+1 = \dfrac{^+1}{^+3} \times {}^+3$. **o.** True, $^+1 < {}^+3$.

q. False, $\overline{AB} \cup \overline{BC} = \overline{AC}$. **s.** True, $^+3 + {}^+5\frac{1}{2} > {}^+5$.

3. **a.** 1. $\overline{AB}$ is a line segment.
2. $\overline{AB} = \overline{AB}$.
3. $m(\overline{AB}) = m(\overline{AB})$.
4. $\overline{AB} \cong \overline{AB}$.

b. 1. $\overline{AB}$ and $\overline{CD}$ are line segments and $\overline{AB} \cong \overline{CD}$.
2. $m(\overline{AB}) = m(\overline{CD})$.
3. $m(\overline{CD}) = m(\overline{AB})$.
4. $\therefore \overline{CD} \cong \overline{AB}$.

c. 1. $\overline{AB}$, $\overline{CD}$, and $\overline{EF}$ are line segments, $\overline{AB} \cong \overline{CD}$, and $\overline{CD} \cong \overline{EF}$.
2. $m(\overline{AB}) = m(\overline{CD})$ and $m(\overline{CD}) = m(\overline{EF})$.
3. $\therefore m(\overline{AB}) = m(\overline{EF})$.
4. $\therefore \overline{AB} \cong \overline{EF}$.

5. 1. $A,B,C,D,E,F \in S$.
2. $A\text{–}B\text{–}C$, $D\text{–}E\text{–}F$, $\overline{AB} \cong \overline{DE}$, and $\overline{BC} \cong \overline{EF}$.
3. $m(\overline{AB}) = m(\overline{DE})$.
$m(\overline{BC}) = m(\overline{EF})$.
4. $m(\overline{AB}) + m(\overline{BC}) = m(\overline{DE}) + m(\overline{EF})$.
5. $m(\overline{AB}) + m(\overline{BC}) = m(\overline{AC})$.
6. $m(\overline{DE}) + m(\overline{EF}) = m(\overline{DF})$.
7. $m(\overline{AC}) = m(\overline{DF})$.
8. $\overline{AC} \cong \overline{DF}$.

7.

8. **a.** False

c. False

e. True

$$A \quad B \quad\quad E \quad F \quad\quad C \quad\quad D$$

$$m(\overline{AB}) = m(\overline{CD}); \therefore m(\overline{AB}) + m(\overline{EF}) = m(\overline{CD}) + m(\overline{EF}).$$

Exercise set 4.4, p. 101

1. **a.** $^{+}4, ^{+}4$ **b.** $^{+}3, ^{+}3$ mules   **c.** $^{+}5, ^{+}5$ tules
   d. $^{+}3, ^{+}3$ mules, $^{+}4, ^{+}4$ tules

2. **a.** (1) $^{+}3.5 \leqslant m(\overline{AB}) < ^{+}4.5.$ (2) $^{+}1$ (3) $^{+}0.5$
     (4) $\dfrac{^{+}0.5}{^{+}4}$

   c. (1) $^{+}4.035 \leqslant m(\overline{AB}) < ^{+}4.045.$ (2) $^{+}0.01$ (3) $^{+}0.005$
     (4) $\dfrac{^{+}0.005}{^{+}4.04}$

3. $m(\overline{CD}) \doteq ^{+}5$ means that the measure of $\overline{CD}$ is given correct to the nearest whole unit [that is, $^{+}4.5 \leqslant m(\overline{CD}) < ^{+}5.5$], whereas $m(\overline{CD}) \doteq ^{+}5.0$ means that the measure of $\overline{CD}$ is given correct to the nearest one-tenth of a whole unit [that is, $^{+}4.95 \leqslant m(\overline{CD}) < ^{+}5.05$].

4. **a.** (1) $^{+}6.605 \leqslant m(\overline{AB}) < ^{+}6.615.$ (2) $^{+}0.01$ (3) $^{+}0.005$
     (4) $\dfrac{^{+}0.005}{^{+}6.61}$

   c. (1) $^{+}2\frac{1}{4} \leqslant m(\overline{AB}) < ^{+}2\frac{3}{4}.$ (2) $\dfrac{^{+}1}{^{+}2}$ (3) $\dfrac{^{+}1}{^{+}4}$
     (4) $\dfrac{^{+}1}{^{+}9}$

   e. (1) $^{+}2\frac{5}{16} \leqslant m(\overline{AB}) < ^{+}2\frac{7}{16}.$ (2) $\dfrac{^{+}1}{^{+}8}$ (3) $\dfrac{^{+}1}{^{+}16}$
     (4) $\dfrac{^{+}1}{^{+}38}$

5. **a.** (1) $^{+}2.45 \times 10^{5}$ (2) $^{+}1000$ (3) $^{+}500$
     (4) $\dfrac{^{+}500}{^{+}2.45 \times 10^{5}} = \dfrac{^{+}1}{^{+}490}.$

   c. (1) $^{+}2.4500 \times 10^{5}$ (2) $^{+}10$ (3) $^{+}5$
     (4) $\dfrac{^{+}5}{^{+}2.4500 \times 10^{5}} = \dfrac{^{+}1}{^{+}49000}.$

ANSWERS AND SUGGESTIONS FOR SELECTED EXERCISES

Exercise set 4.5, p. 109

2. a. $|^+10 - 0| = ^+10.$ **c.** False **e.** False **g.** $^+45$
   i. True **k.** $^+80$ **m.** False **o.** True

3. For all rays AB, AC, and AD, $\overrightarrow{AC}$ is between $\overrightarrow{AB}$ and $\overrightarrow{AD}$ if and only if

  1. $\overrightarrow{AB}$, $\overrightarrow{AC}$, and $\overrightarrow{AD}$ are coplanar, and
  2. $m \circ (\angle BAC) + m \circ (\angle CAD) = m \circ (\angle BAD).$

5. a. No. Although $\angle ABC \neq \angle DEF$, $m \circ (\angle ABC)$ may be equal to $m \circ (\angle DEF)$. Therefore, $\angle ABC$ may be congruent to $\angle DEF$.
   b. Yes. $\angle ABC \cong \angle DEF$ if and only if $m \circ (\angle ABC) = m \circ (\angle DEF).$
   c. Yes. Since $m \circ (\angle ABC) > m \circ (\angle DEF)$, $m \circ (\angle ABC) \neq m \circ (\angle DEF)$. If $m \circ (\angle ABC) \neq m \circ (\angle DEF)$, then $\angle ABC \not\cong \angle DEF$. $\therefore$ If $\angle ABC \neq \angle DEF$, then $\angle ABC \neq \angle DEF.$

6. a. No

b. Yes

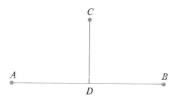

c. No

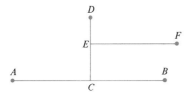

Exercise set 4.6, p. 113

1. **a.** $\angle FEB$ and $\angle CBE$, $\angle DEB$ and $\angle ABE$
    b. $\angle ABE$ and $\angle BEF$, $\angle DEB$ and $\angle CBE$
    c. $\angle HED$ and $\angle ABE$, $\angle DEB$ and $\angle ABG$, $\angle GBC$ and $\angle BEF$, $\angle HEF$ and $\angle HBC$

5. **a.** Prove: If two angles of a linear pair are congruent, then they are right angles.

    1. $\angle ABC$ and $\angle CBD$ are a linear pair, $\angle ABC \cong \angle CBD$.
    2. $\overrightarrow{BA} \cup \overrightarrow{BD} = \overleftrightarrow{AD}$.
    3. $\angle ABC$ and $\angle CBD$ are supplementary.
    4. $m_\circ(\angle ABC) + m_\circ(\angle CBD) = {}^+180$.
    5. $m_\circ(\angle ABC) = m_\circ(\angle CBD)$.
    6. $\therefore m_\circ(\angle ABC) + m_\circ(\angle CBD) = {}^+180$.
    7. $\therefore m_\circ(\angle ABC) = {}^+90$.
    8. $\therefore \angle ABC$ is a right angle.
    9. $m_\circ(\angle CBD) + m_\circ(\angle CBD) = {}^+180$.
    10. $\therefore m_\circ(\angle CBD) = {}^+90$.
    11. $\therefore \angle CBD$ is a right angle.

The proof that "If two angles of a linear pair are right angles, then they are congruent" is left to the reader.

7. **1.** $\overleftrightarrow{AE}$, $\overleftrightarrow{CD}$, transversal $\overleftrightarrow{CB}$, $\angle ABC$ and $\angle DCB$ are alternate interior angles, and $\angle ABC \cong \angle DCB$.

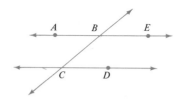

    2. $\angle ABC$ and $\angle CBE$ are adjacent angles.
    3. $\angle ABC$ and $\angle CBE$ are a linear pair.
    4. $\angle ABC$ and $\angle CBE$ are supplementary angles.
    5. $m_\circ(\angle ABC) + m_\circ(\angle CBE) = {}^+180$.
    6. $m_\circ(\angle ABC) = m_\circ(\angle DCB)$.

7. $\therefore m_\circ(\angle DCB) + m_\circ(\angle CBE) = {}^+180.$

8. $\therefore \angle DCB$ and $\angle CBE$ are supplementary angles.

Exercise set 5.1, p. 127

1. **a.** None **c.** Simple
e. Simple, closed, simple closed, polygon, convex curve
g. Closed
i. Simple, closed, simple closed, polygon, convex curve
k. Simple, closed, simple closed, convex curve
m. Simple, closed, simple closed, polygon

2. **a.** False **c.** True **e.** True

3. **a.** $\overrightarrow{AB}$ crosses curve $\mathscr{C}$ at D because the interior of every circle with center at D contains points of the curve in both of the opposite half-planes of $\overleftrightarrow{AB}$.

c. $\overrightarrow{AB}$ does not cross curve $\mathscr{C}$ at D because the interior of every circle with center at D contains points of the curve in both of the opposite half-planes of $\overleftrightarrow{AB}$.

e. $\overrightarrow{AB}$ crosses curve $\mathscr{C}$ at D. Same reason as part a.

4. **a.** P, exterior **c.** P, interior **e.** Q, interior
  Q, interior Q, exterior P, interior

Exercise set 5.2, p. 131

1. **a.** $\dfrac{{}^+31}{{}^+4}$

2. **a.** ${}^+2\pi$ **c.** ${}^+8\pi$

3. **a.** ${}^+6.28$ **c.** ${}^+25.12$

4. **a.** (1) ${}^+9$ **b.** (1) ${}^+10$

Exercise set 5.3, p. 136

1. **A.** **a.** ${}^+3$ **B.** **a.** (1) ${}^+2$
  b. $\dfrac{{}^+1}{{}^+3}$ (2) ${}^+2$
      (3) ${}^+5$
  c. ${}^+3, \dfrac{{}^+1}{{}^+3}$ (4) ${}^+5$

answers and suggestions for selected exercises

2. a.

Line segment	Approximate measure	g.p.e.	r.e.
$\overline{AB}$		$^+0.05$	$\dfrac{^+1}{^+70}$
$\overline{BC}$		$^+0.05$	$\dfrac{^+1}{^+134}$
$\overline{CD}$		$^+0.05$	$\dfrac{^+1}{^+248}$
$\overline{DE}$		$^+0.05$	$\dfrac{^+1}{^+167}$
$\overline{EF}$		$^+0.05$	$\dfrac{^+1}{^+184}$
$\mathscr{C}$	$^+40.1$	$^+0.25$	$\dfrac{^+10}{^+1604}$

Exercise set 6.1, p. 142

1. In the diagram, P is the set of polygons, Q is the set of quadrilaterals, T is the set of trapezoids, P_1 is the set of parallelograms, R is the set of rectangles, R_1 is the set of rhombuses,
and S is the set of squares.

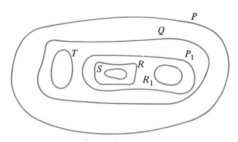

4. **a.** A *pentagon* is a polygon which is the union of five line
segments.
c. An *octagon* is a polygon which is the union of eight line
segments.
5. **a.** $\overline{AB}, \overline{BC}, \overline{CD}, \overline{DE}, \overline{EA}$ **b.** B and D
c. $\angle EAB, \angle ABC, \angle BCD, \angle CDE, \angle DEA, \angle EAB$
d. Yes, for every $X, Y \in \mathscr{C}$, $\overline{XY} \subseteq \mathscr{C} \cup \mathrm{Int}(\mathscr{C})$.

e. Yes, definition of a diagonal   **f.** $\overline{AD}, \overline{BE}, \overline{BD}, \overline{CE}$

g. Five   **h.** Pentagon

7. c.

V_1	V_2	V_3	V_4	V_5	Total
2	2	1	0	0	5

d.

V_1	V_2	V_3	V_4	V_5	V_6	Total
3	3	2	1	0	0	9

e.

V_1	V_2	V_3	V_4	V_5	V_6	V_7	Total
4	4	3	2	1	0	0	14

f.

Number of sides:	3	4	5	6	7	8	9	10	11
Number of diagonals:	0	2	5	9	14	20	27	35	44

h. Yes

Exercise set 6.2, p. 146

1. a. (1) $\overline{DE}$ (2) $\overline{AC}$ (3) $\overline{BC}$ and $\overline{EF}$

(4) $\angle D$ (5) $\angle B$ (6) $\angle C$ and $\angle F$.

2. $ABC \leftrightarrow ACB$, $\overline{AB}$ and $\overline{AC}$, $\overline{BC}$ and $\overline{CB}$, $\overline{AC}$ and $\overline{AB}$, $\angle A$ and $\angle A$, $\angle B$ and $\angle C$, $\angle C$ and $\angle B$

$ABC \leftrightarrow BCA$, $\overline{AB}$ and $\overline{BC}$, $\overline{AC}$ and $\overline{BA}$, $\overline{BC}$ and $\overline{CA}$, $\angle A$ and $\angle B$, $\angle B$ and $\angle C$, $\angle C$ and $\angle A$

$ABC \leftrightarrow CAB$, $\overline{AB}$ and $\overline{CA}$, $\overline{AC}$ and $\overline{CB}$, $\overline{BC}$ and $\overline{AB}$, $\angle A$ and $\angle C$, $\angle B$ and $\angle A$, $\angle C$ and $\angle B$

$ABC \leftrightarrow CBA$, $\overline{AB}$ and $\overline{CB}$, $\overline{AC}$ and $\overline{CA}$, $\overline{BC}$ and $\overline{BA}$, $\angle A$ and $\angle C$, $\angle B$ and $\angle B$, $\angle C$ and $\angle A$

Exercise set 6.3, p. 148

1. A. a. Not congruent

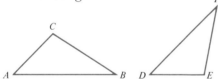

b. Not congruent

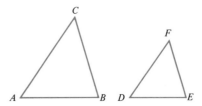

c. Not congruent

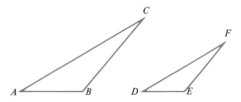

C. **a.** Not congruent

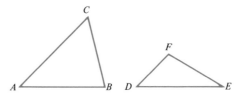

b. Not congruent

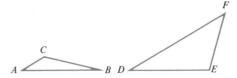

c. Congruent

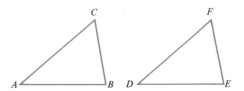

Exercise set 6.4, p. 149

1. **a.** Not congruent   **c.** Not congruent   **e.** Not congruent
    g. Congruent, $ABC \leftrightarrow ABC$, ASA triangle congruence postulate

3. **a.** Yes   **b.** $ABD \leftrightarrow CBD$, SSS triangle congruence postulate

5. **a.** No

7. **a.** Yes   **b.** $QTR \leftrightarrow SRT$, ASA triangle congruence postulate

Exercise set 6.5, p. 154

1. **1.** $\triangle ABC$ and $\overline{AB} \cong \overline{CB}$.
    2. $\overline{AB} \cong \overline{BA}$ and $\overline{BC} \cong \overline{CB}$.
    3. $\therefore \overline{BC} \cong \overline{BA}$.
    4. $\overline{AC} \cong \overline{CA}$.
    5. $\therefore \triangle ABC \cong \triangle CBA$.
    6. $\therefore \angle A \cong \angle C$.
    7. $\therefore \angle C \cong \angle A$.

3. **1.** $\triangle ABC$ is equilateral.
    2. $\overline{AB} \cong \overline{BC}, \overline{AC} \cong \overline{BC}, \overline{AB} \cong \overline{AC}$.
    3. $\angle A \cong \angle C, \angle B \cong \angle A, \angle C \cong \angle B$.
    4. $\therefore \triangle ABC$ is equiangular.

    (*Note:* Hereafter we denote such a combination of congruences by $\overline{AB} \cong \overline{BC} \cong \overline{AC}$.)

5. **1.** $ABCD$ is a parallelogram.
    2. $\overline{AB} \parallel \overline{CD}$ and $\overline{AB} \cong \overline{CD}$.
    3. $\overline{BD}$ is a diagonal.
    4. $\therefore \triangle CDB \cong \triangle ABD$.
    5. $\therefore \overline{BC} \cong \overline{DA}$.
    6. $\therefore \angle ADB \cong \angle CBD$.
    7. $\overleftrightarrow{BD}$ is a transversal.
    8. $\therefore \overleftrightarrow{DA} \parallel \overleftrightarrow{BC}$.
    9. $\therefore \overline{DA} \parallel \overline{BC}$.

answers and suggestions for selected exercises

7.

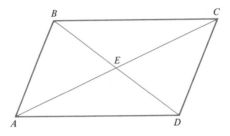

1. $ABCD$ is a parallelogram.
2. $\overline{AC}$ and $\overline{BD}$ are diagonals.
3. $\overline{AB} \cong \overline{CD}$ and $\overline{AB} \parallel \overline{CD}$.
4. $\overleftrightarrow{AB} \parallel \overleftrightarrow{CD}$.
5. $\angle ABE \cong \angle CDE$.
6. $\angle EAB \cong \angle ECD$.
7. $\therefore \triangle ABE \cong \triangle CDE$.
8. $\therefore \overline{AE} \cong \overline{CE}$.
9. $\therefore E$ bisects $\overline{AC}$.
10. $\therefore \overline{BE} \cong \overline{DE}$.
11. $\therefore E$ bisects $\overline{BD}$.

9.
1. $ABCD$ is a rectangle.
   $\overline{AC}$ and $\overline{BD}$ are diagonals.
2. $ABCD$ is a parallelogram.
3. $\overline{DC} \cong \overline{AB}$.
   $\overline{BC} \cong \overline{CB}$.
4. $\angle ABC$ is a right angle.
   $\angle DCB$ is a right angle.
5. $m_\circ(\angle ABC) = {}^+90$.
   $m_\circ(\angle DCB) = {}^+90$.
6. $m_\circ(\angle ABC) = m_\circ(\angle DCB)$.
7. $\angle ABC \cong \angle DCB$.
8. $\triangle ABC \cong \triangle DCB$.
9. $\therefore \overline{AC} \cong \overline{DB}$.

10.
1. $ABCD$ is a rhombus and $\angle ABC$ is a right angle.
2. $ABCD$ is a parallelogram.
3. $\therefore ABCD$ is a rectangle.
4. $\therefore \overline{AB} \cong \overline{BC} \cong \overline{CD} \cong \overline{DA}$.
5. $\therefore ABCD$ is a square.

ANSWERS AND SUGGESTIONS FOR SELECTED EXERCISES

Exercise set 6.6, p. 158

1. **a.** Not similar   **c.** Not similar
    e. Similar, $ABC \leftrightarrow STU$, SAS triangle similarity postulate
    g. Similar, $DEF \leftrightarrow MNO$, SAS triangle similarity postulate

3. 1. $\triangle ABC, \overleftrightarrow{DE}$ such that $\overleftrightarrow{DE}$ bisects $\overline{BA}, \overleftrightarrow{DE} \parallel \overleftrightarrow{AC}, \overline{AB} \cap \overleftrightarrow{DE} = \{D\}$, and $\overline{BC} \cap \overleftrightarrow{DE} = \{E\}$.
    2. $\overleftrightarrow{DE} \parallel \overleftrightarrow{AC}$.
    3. $\therefore \angle A \cong \angle EDB$.
    4. $\angle B \cong \angle B$.
    5. $\therefore \triangle ABC \sim \triangle DBE$.
    6. $\therefore \dfrac{m(\overline{AB})}{m(\overline{DB})} = \dfrac{m(\overline{CB})}{m(\overline{EB})}$.
    7. $\dfrac{m(\overline{AD}) + m(\overline{DB})}{m(\overline{DB})} = \dfrac{m(\overline{CE}) + m(\overline{EB})}{m(\overline{EB})}$.
    8. $\dfrac{m(\overline{AD})}{m(\overline{DB})} + \dfrac{m(\overline{DB})}{m(\overline{DB})} = \dfrac{m(\overline{CE})}{m(\overline{EB})} + \dfrac{m(\overline{EB})}{m(\overline{EB})}$.
    9. $m(\overline{EB}) = m(\overline{CE})$.   (Several steps)
    10. $\overleftrightarrow{DE}$ bisects $\overline{BC}$.   (Several steps)

5. 1,7; 2,6; 3,10; 8,9; 11,12.

Exercise set 6.7, p. 165

1.

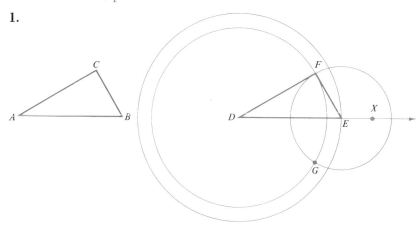

Steps in the construction:

1. "Draw a circle" with center at D and radius $m(\overline{AB})$. The circle will intersect $\overrightarrow{DX}$ in exactly one point E.

2. "Draw a circle" with center at E and radius $m(\overline{BC})$.
3. "Draw a circle" with center at A and radius $m(\overline{AC})$.
4. The circle with center at E will intersect the circle with center at D in exactly two points F and G.
5. "Draw line segments DF and EF."
6. $\triangle DEF \cong \triangle ABC$.

Justification:

By our construction $\overline{AB} \cong \overline{DE}$, $\overline{BC} \cong \overline{EF}$, and $\overline{AC} \cong \overline{DE}$; therefore, by the SSS triangle congruence postulate, $\triangle DEF \cong \triangle ABC$.

Exercise set 7.1, p. 168

1. **A.**   **a.** True   **c.** True   **e.** $\overline{AC}$ **g.** $\overline{EF} \setminus \{E,F\}$
     i. True   **k.** $\overleftrightarrow{AI} \cup \overleftrightarrow{CG}$ **m.** False   **o.** False
     q. True

   B.   **a.**    **c.**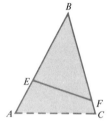

2.   **a.** 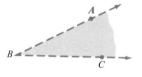   **c.**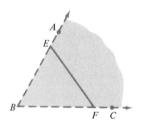

   e. 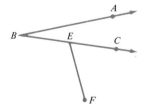   **g.**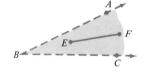

i.

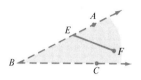

k.

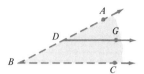

m.

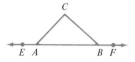

o.

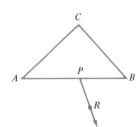

q.

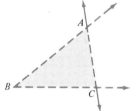

s.

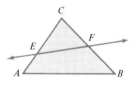

u.

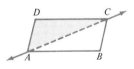

w.

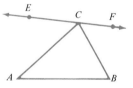

y.

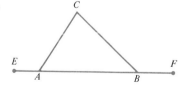

Exercise set 7.2, p. 174

3. **a.** ⁺12   **b.** ⁺1   **c.** ⁺27

Exercise set 7.3, p. 179

1. **a.** (1) ⁺3.35, ⁺3.45   (2) ⁺5.55, ⁺5.65   (3) ⁺18.0
  (4) ⁺17.80, ⁺18.20

(5) (1) $^+0.005, \dfrac{^+1}{^+68}$ (2) $^+0.05, \dfrac{^+1}{^+112}$ (4) $^+0.20, \dfrac{^+1}{^+90}$

(6) $^+18$ (7) $^+19.04$ (8) $^+18.5925, {}^+19.4925$

(9)

	g.p.e.	r.e.
(1)	$^+0.05$	$\dfrac{^+1}{^+68}$
(2)	$^+0.05$	$\dfrac{^+1}{^+112}$
(8)	$^+0.45$	$\dfrac{^+45}{^+1904}$

Note that $\dfrac{^+1}{^+68} + \dfrac{^+1}{^+112} = \dfrac{^+45}{^+1904}$ and $^+0.45 = {}^+0.05 \times {}^+3.4 + {}^+0.05 \times {}^+5.6$.

(10) $^+19$

Exercise set 7.4, p. 185

1. 1. $m(\blacktriangle ABD) = \dfrac{^+1}{^+2} \times m(\overline{AD}) \times m(\overline{BD})$.

2. $m(\blacktriangle BCD) = \dfrac{^+1}{^+2} \times m(\overline{CD}) \times m(\overline{BD})$.

3. $m(\blacktriangle ABC) + m(\blacktriangle BCD) = m(\blacktriangle ABD)$.

4. $\therefore m(\blacktriangle ABC) = m(\blacktriangle ABD) - m(\blacktriangle BCD)$.

5. $\therefore m(\blacktriangle ABC) = \dfrac{^+1}{^+2} \times m(\overline{AD}) \times m(\overline{BD})$

$$-\dfrac{^+1}{^+2} \times m(\overline{CD}) \times m(\overline{BD}).$$

6. $\therefore m(\blacktriangle ABC) = \dfrac{^+1}{^+2} \times m(\overline{BD}) \times (m(\overline{AD}) - m(\overline{CD}))$.

7. $\therefore m(\blacktriangle ABC) = \dfrac{^+1}{^+2} \times m(\overline{BD}) \times m(\overline{AC})$.

2. $^+3$

3. $^+9$

5. $m(\overline{EH}) = \dfrac{^+1}{^+2} \times m(\overline{CG})$.

7. **a.** $^+45$ **b.** (1) $^+34$ (2) $^+8$ (3) $^+6$
   c. (1) $^+16$ (2) $^+4$

9. **a.** $^+12$ **b.** $^+6$ **c.** $^+6$ **d.** $^+10$ **e.** $^+22$

11. $^+40$

Exercise set 8.1, p. 196

1. **a.** $^+17$ **b.** $^+2$ **c.** $^+34$ **d.** Yes

3. **a.** Trapezoidal   **b.** Base   **c.** Lateral face
   d. Lateral edge   **e.** Right prism

5. $^+12$

7. **1.** $\mathscr{F}$ is a cube.
   2. e is the measure of a lateral edge.
   3. The measure of the boundary of the base is ^+4e.
   4. $\therefore$ The measure of the lateral surface of the cube is $^+4e \times e$ or $^+4e^2$.

10. $^+160$

Exercise set 8.2, p. 201

1. **a.** $^+42\pi, ^+60\pi$ **c.** $^+42\pi, ^+140\pi$ **e.** $^+80\pi, ^+112\pi$

3. The measure of the lateral surface and total measure each is four and nine times the original measures when the radii of the boundaries of the bases and the measure of an altitude are doubled and tripled, respectively.

4. **a.** $^+10\pi, ^+14\pi$ **c.** $^+90\pi, ^+126\pi$

7. Yes

10. (*Hint*: See Definition 8.6.)

Exercise set 8.3, p. 206

1. **a.** $^+24$ **c.** $^+96$

2. If the measure of one edge of a solid rectangular parallelepiped $\mathscr{S}$ is doubled, then $m(\mathscr{S})$ is doubled.

5. **a.** $^+8$

7. If the measure of the edges of a cubic solid $\mathscr{S}$ are doubled, then $m(\mathscr{S})$ is eight times the original $m(\mathscr{S})$.

Exercise set 8.4, p. 210

1. a. $^{+}42$

2. a. $^{+}25$

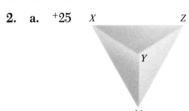

3. a. $^{+}36\pi$ c. $^{+}324\pi$ e. $^{+}72\pi$

   g. If the measure of the altitude of a solid right circular cylinder $\mathscr{S}$ is doubled, then $m(\mathscr{S})$ is doubled. If the measure of the altitude of a solid right circular cylinder $\mathscr{S}$ is tripled, then $m(\mathscr{S})$ is tripled.

4. a. $^{+}4\pi$ c. $^{+}12\pi$ e. $^{+}16\pi$
   g. If the measure of the radius of the base of a solid right circular cone $\mathscr{S}$ is doubled, then $m(\mathscr{S})$ is four times the original $m(\mathscr{S})$. If the measure of the radius of the base of a solid right circular cone $\mathscr{S}$ is tripled, then $m(\mathscr{S})$ is nine times the original $m(\mathscr{S})$.

5.

Measure of each edge of the cube	M_1: total measure of the cube	M_2: measure of solid cube	Ratio: $\dfrac{M_1}{M_2}$
$^{+}1$	$^{+}6$	$^{+}1$	$^{+}6$
$^{+}2$	$^{+}24$	$^{+}8$	$^{+}3$
$^{+}3$	$^{+}54$	$^{+}27$	$^{+}2$
$^{+}4$	$^{+}96$	$^{+}64$	$^{+}1\dfrac{1}{2}$
$^{+}5$	$^{+}150$	$^{+}125$	$^{+}1\dfrac{1}{5}$
$^{+}6$	$^{+}216$	$^{+}216$	$^{+}1$
$\vdots$	$\vdots$	$\vdots$	$\vdots$
x	$^{+}6x^2$	x^3	$\dfrac{^{+}6}{x}$

Exercise set 9.1, p. 218

1.　**a.** Yes　　**b.** Yes　　**c.** Yes
　　d. (1) $b * e = c$.　　(2) $[b * e] * d = c * d = b$.
　　　　(3) $[a * c] * a = c * a = c$.
　　e. $(a * b) * c = a$, $a * (b * a) = a$.　No
　　f. $(a * c) * d = b$, $a * (c * d) = b$.　No
　　g. $(b * c) * d = d$, $b * (c * d) = c$.　Yes
　　h. No.　$d * a \neq a * d$.
　　i. (1)　c　　(2)　c　　(3)　$b, d,$ or e

3.　**a.** $a \cdot b = b$.　　**b.** $a \odot b = c$.　　**c.** $a \circledcirc c = c$.
　　　　　　　　　　$b \odot c = c$.　　　　$a \circledcirc d = d$.
　　　　　　　　　　$c \odot a = b$.　　　　$b \circledcirc a = b$.
　　d. $c \circledcirc b = a$.
　　　　$c \circledcirc d = d$.
　　　　$d \circledcirc b = c$.

5.　**a.** Yes　　**b.** Yes　　**c.** Yes

7.　**a.**

⊕	0	1	2	3
0	0	1	2	3
1	1	2	3	0
2	2	3	0	1
3	3	0	1	2

b. Yes　　**c.** Yes
d. Yes　　**e.** Yes
g. Yes

　　h. $2 \bullet (1 \oplus 3) = 2 \bullet 0 = 0$.　　**i.** $3 \oplus (2 \bullet 1) = 1$.
　　　　$(2 \bullet 1) \oplus (2 \bullet 3) = 2 \oplus 2 = 0$.　　　$(3 \oplus 2) \bullet (3 \oplus 1) = 0$.
　　　　No　　　　　　　　　　　　　　　　Yes.　⊕ does not dis-
　　　　　　　　　　　　　　　　　　　　　tribute over •.

9.　**a.** T (Commutative property of $*$)
　　c. T (Associative property of $*$)
　　e. T (Associative property of #)
　　g. F (# does not distribute over $*$)
　　i. T (Closure property of # and commutative property of $*$)

Exercise set 9.2, p. 223

1. **a.** Yes. $a \cdot a = a$; $b \cdot a = b$ and $a \cdot b = b$; $c \cdot a = c$ and $a \cdot c = c$. Therefore, a is the identity with respect to $\cdot$.

b. Yes. The inverse of a is a; the inverse of b is c; the inverse of c is b.

c. Yes

d. Yes. $a \cdot a = a \cdot a$; $b \cdot a = a \cdot b$; $c \cdot a = a \cdot c$; $b \cdot c = c \cdot b$; $b \cdot b = b \cdot b$; $c \cdot c = c \cdot c$.

3. **a.**

⊛	1	2	3	4	5	6	7	8	9	10	11	12
1	2	3	4	5	6	7	8	9	10	11	12	1
2	3	4	5	6	7	8	9	10	11	12	1	2
3	4	5	6	7	8	9	10	11	12	1	2	3
4	5	6	7	8	9	10	11	12	1	2	3	4
5	6	7	8	9	10	11	12	1	2	3	4	5
6	7	8	9	10	11	12	1	2	3	4	5	6
7	8	9	10	11	12	1	2	3	4	5	6	7
8	9	10	11	12	1	2	3	4	5	6	7	8
9	10	11	12	1	2	3	4	5	6	7	8	9
10	11	12	1	2	3	4	5	6	7	8	9	10
11	12	1	2	3	4	5	6	7	8	9	10	11
12	1	2	3	4	5	6	7	8	9	10	11	12

b. Yes   **c.** Yes   **d.** Yes

e. The identity with respect to ⊛ is 12. $1 \circledast 12 = 1$ and $12 \circledast 1 = 1$; $2 \circledast 12 = 2$ and $12 \circledast 2 = 2$; $12 \circledast 3 = 3$ and $3 \circledast 12 = 3$; and so on.

f. $1 \circledast 11 = 12$ and $11 \circledast 1 = 12$;
$2 \circledast 10 = 12$ and $10 \circledast 2 = 12$;

$3 \circledast 9 = 12$ and $9 \circledast 3 = 12$;
$4 \circledast 8 = 12$ and $8 \circledast 4 = 12$;
$5 \circledast 7 = 12$ and $7 \circledast 5 = 12$;
$6 \circledast 6 = 12$.

g. Yes   h. Yes   i. Yes

Exercise set 9.3, p. 226

1. a. Relation, function, into, one-to-one
    b. Relation, function, one-to-one, onto, into, transformation
    c. Relation, function

2. a. Onto, one-to-one, into, transformation
    c. None

3. a. 4, 7, 6, 5   b. 4, 6   e. {4,6}   f. Yes

Exercise set 9.4, p. 231

1. a.

$\circ$	t_1	t_2	t_3	t_4	t_5	t_6
t_1	t_1	t_2	t_3	t_4	t_5	t_6
t_2	t_2	t_1	t_5	t_6	t_3	t_4
t_3	t_3	t_4	t_1	t_2	t_6	t_5
t_4	t_4	t_3	t_6	t_5	t_1	t_2
t_5	t_5	t_6	t_2	t_1	t_4	t_3
t_6	t_6	t_5	t_4	t_3	t_2	t_1

b. Yes. t_1 is the identity with respect to $\circ$ in T_M.
    $t_1 \circ t_1 = t_1$; $t_2 \circ t_1 = t_2$ and $t_1 \circ t_2 = t_2$, $t_3 \circ t_1 = t_3$ and $t_1 \circ t_3 = t_3$; and so on.

c. Yes. The inverse of t_1 is t_1; the inverse of t_2 is t_2; the inverse of t_3 is t_3; the inverse of t_4 is t_5; the inverse of t_5 is t_4; the inverse of t_6 is t_6.

d. $t_2 \circ t_3 \neq t_3 \circ t_2$.   e. (1) $t_4 \circ t_3 = t_6$.   f. $6 \cdot 6 \cdot 6 = 216$.
                                               (2) $t_2 \circ t_4 = t_6$.
                                               (3) No

g. Yes. The system T_M, $\circ$ has properties G_1, G_2, and G_3.
h. No. $t_2 \circ t_3 \neq t_3 \circ t_2$.

Exercise set 9.5, p. 235

1. a. 1. $t_1, t_2 \in T_S$.
   2. $t_1^{\,-1} \circ (t_1 \circ t_2) = (t_1^{\,-1} \circ t_1) \circ t_2$
   3. $\qquad\qquad = i \circ t_2$
   4. $\qquad\qquad = t_2$.
   5. $\therefore t_1^{\,-1} \circ (t_1 \circ t_2) = t_2$.

   c. 1. $t_1, t_2 \in T_S$.
   2. $(t_1 \circ t_2)^{-1} \circ (t_1 \circ t_2) = i$.
   3. $(t_2^{\,-1} \circ t_1^{\,-1}) \circ (t_1 \circ t_2) = \left[(t_2^{\,-1} \circ t_1^{\,-1}) \circ t_1 \right] \circ t_2$
   4. $\qquad\qquad\qquad\qquad = \left[t_2^{\,-1} \circ (t_1^{\,-1} \circ t_1) \right] \circ t_2$
   5. $\qquad\qquad\qquad\qquad = (t_2^{\,-1} \circ i) \circ t_2$
   6. $\qquad\qquad\qquad\qquad = t_2^{\,-1} \circ t_2$
   7. $\qquad\qquad\qquad\qquad = i$.
   8. $\therefore (t_2^{\,-1} \circ t_1^{\,-1}) \circ (t_1 \circ t_2) = i$.
   9. $\therefore (t_2 \circ t_1)^{-1} = t_2^{\,-1} \circ t_1^{\,-1}$.

Exercise set 10.2, p. 245

1. 1. t is a translation of a plane α, A, B, and C are points in α, and A–B–C.
   2. $\therefore AB + BC = AC$.
   3. $AB = A'B'$, $BC = B'C'$, and $AC = A'C'$.
   4. $\therefore A'B' + B'C' = A'C'$.
   5. $\therefore A'$–B'–C'.

3. 1. t is a translation of a plane α, $\overline{AB}$ and $\overline{CD}$ are in α, and $\overline{AB} \parallel \overline{CD}$.
   2. $\overline{AA'} \parallel \mathscr{L}$, $\overline{BB'} \parallel \mathscr{L}$, $\overline{CC'} \parallel \mathscr{L}$, and $\overline{DD'} \parallel \mathscr{L}$.
   3. $\overline{AA'} \cong \overline{BB'}$ and $\overline{CC'} \cong \overline{DD'}$.
   4. $\overline{AA'} \parallel \overline{BB'}$ and $\overline{CC'} \parallel \overline{DD'}$.
   5. $\therefore AA'B'B$ and $CC'D'D$ are parallelograms.
   6. $\therefore \overline{AB} \parallel \overline{A'B'}$ and $\overline{CD} \parallel \overline{C'D'}$.
   7. $\therefore \overline{A'B'} \parallel \overline{AB}$.
   8. $\therefore \overline{A'B'} \parallel \overline{CD}$.
   9. $\therefore \overline{A'B'} \parallel \overline{C'D'}$.

5. 1. t is a translation of a plane α, $A, B, C \in \alpha$, and A, B, and C are noncollinear.
   2. $AB + BC > AC$.
   3. $AB = A'B'$, $BC = B'C'$, and $AC = A'C'$.
   4. $\therefore A'B' + B'C' > A'C'$.
   5. $\therefore A'$, B', and C' are noncollinear.

7. a. $t_1 \circ t_2$.

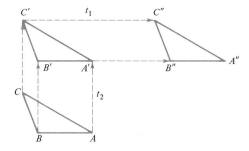

$t_2 \circ t_1$.

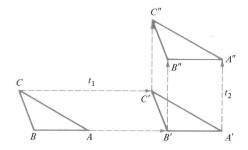

Exercise set 10.3, p. 254

1.

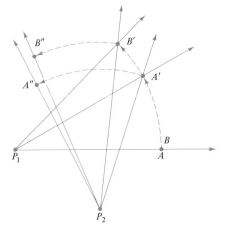

$m_\circ(\angle AP_1A') = m_\circ(\angle B'P_2B'') = {}^+30.$
$m_\circ(\angle BP_1B') = m_\circ(\angle A'P_2A'') = {}^+45.$
$r_{45} \circ r_{30} \neq r_{30} \circ r_{45}$ since $A = B$, but $A'' \neq B''$.

answers and suggestions for selected exercises

3.

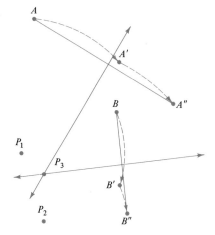

1. *A* and *B* are mapped to *A′* and *B′* by the rotation r_{45} with center P_1.
2. *A′* and *B′* are mapped to *A″* and *B″* by the rotation r_{30} with center P_2.
3. The perpendicular bisectors of $\overline{AA}''$ and $\overline{BB}''$ will intersect in point P_3.
4. P_3 is the center for a rotation that will map *A* and *B* to *A″* and *B″*, respectively.

6. 1. $\angle ABC$ is in plane α and r is a rotation.
2. $AB = A'B'$, $BC = B'C'$, and $CA = C'A'$.
3. $\overline{AB} \cong \overline{A'B'}$, $\overline{BC} \cong \overline{B'C'}$, and $\overline{CA} \cong \overline{C'A'}$.
4. $\therefore \triangle ABC \cong \triangle A'B'C'$.
5. $\therefore \angle ABC \cong \angle A'B'C'$.

Exercise set 10.4, p. 261

1. a.

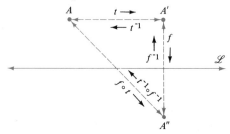

b.

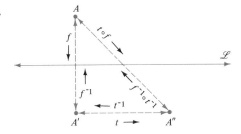

Exercise set 10.5, p. 264

1. **a.** One turning symmetry
 c. One turning symmetry
 e. One folding symmetry
 g. One folding symmetry
 i. Three turning symmetries and four folding symmetries
 k. Two turning symmetries and three folding symmetries

Exercise set 10.6, p. 269

1. **a.**

$\circ$	r_0	r_{120}	r_{240}	f_1	f_2	f_3
r_0	r_0	r_{120}	r_{240}	f_1	f_2	f_3
r_{120}	r_{120}	r_{240}	r_0	f_2	f_3	f_1
r_{240}	r_{240}	r_0	r_{120}	f_3	f_1	f_2
f_1	f_1	f_3	f_2	r_0	r_{240}	r_{120}
f_2	f_2	f_1	f_3	r_{120}	r_0	r_{240}
f_3	f_3	f_2	f_1	r_{240}	r_{120}	r_0

 b. Yes **c.** Yes
 d. Yes. S_3 is closed with respect to $\circ$ and $\circ$ is well defined on S_3.
 e. $r_0 \circ r_0 = r_0$.
  $r_0 \circ r_{120} = r_{120}$ and $r_{120} \circ r_0 = r_{120}$.

answers and suggestions for selected exercises

$$r_0 \circ r_{240} = r_{240} \text{ and } r_{240} \circ r_0 = r_{240}.$$
$$r_0 \circ f_1 = f_1 \text{ and } f_1 \circ r_0 = f_1.$$
$$r_0 \circ f_2 = f_2 \text{ and } f_2 \circ r_0 = f_2.$$
$$r_0 \circ f_3 = f_3 \text{ and } f_3 \circ r_0 = f_3.$$

f. $r_0 \circ r_0 = r_0.$ $f_1 \circ f_1 = r_0.$

$r_{120} \circ r_{240} = r_0.$ $f_2 \circ f_2 = r_0.$

$r_{240} \circ r_{120} = r_0.$ $f_3 \circ f_3 = r_0.$

g. Yes   **h.** Yes   **i.** No. $f_1 \circ f_2 \neq f_2 \circ f_1.$

index

Vertex (*continued*)
  of pyramid, 195
Vertices
  consecutive, of polygon, 138
  opposite, of quadrilateral, 141
  of polygon, 138
  of polyhedron, 191

Volume, 205

Well-defined property
  of binary operations, 216
  of function, 34
Well-defined sets, 3